Other titles in this series

Island Ecology	M. Gorman
Plant-Atmosphere Relations	John Grace
Insect Herbivory	I.D. Hodkinson and M.K. Hughes
Modelling	John N.R. Jeffers
Vegetation Dynamics	John Miles

Outline

Editors

George M. Dunnet
Regius Professor of Natural Histo
University of Aberdeen

Charles H. Gimingham
Professor of Botany,
University of Aberdeen

Editors' Foreword

Both in its theoretical and applied aspects, ecology is developing rapidly. In part because it offers a relatively new and fresh approach to biological enquiry, but it also stems from the revolution in public attitudes towards the quality of the human environment and the conservation of nature. There are today more professional ecologists than ever before, and the number of students seeking courses in ecology remains high. In schools as well as universities the teaching of ecology is now widely accepted as an essential component of biological education, but it is only within the past quarter of a century that this has come about. In the same period, the journals devoted to publication of ecological research have expanded in number and size, and books on aspects of ecology appear in ever-increasing numbers.

These are indications of a healthy and vigorous condition, which is satisfactory not only in regard to the progress of biological science but also because of the vital importance of ecological understanding to the well-being of man. However, such rapid advances bring their problems. The subject develops so rapidly in scope, depth and relevance that textbooks, or parts of them, soon become out-of-date or inappropriate for particular courses. The very width of the front across which the ecological approach is being applied to biological and environmental questions introduces difficulties: every teacher handles his subject in a different way and no two courses are identical in content.

This diversity, though stimulating and profitable, has the effect that no single text-book is likely to satisfy fully the needs of the student attending a course in ecology. Very often extracts from a wide range of books must be consulted, and while this may do no harm it is time-consuming and expensive. The present series has been designed to offer quite a large number of relatively small booklets, each on a restricted topic of fundamental importance which is likely to constitute a self-contained component of more comprehensive courses. A selection can then be made, at reasonable cost, of texts appropriate to particular courses or the interests of the reader. Each is written by an acknowledged expert in the subject, and is intended to offer an up-to-date, concise summary which will be of value to those engaged in teaching, research or applied ecology as well as to students.

Studies in Ecology

Animal Population Dynamics

ROBERT MOSS
ADAM WATSON
Institute of Terrestrial Ecology
Banchory, Kincardineshire
Scotland

JOHN OLLASON
Department of Zoology
University of Aberdeen
Scotland

CHAPMAN AND HALL
LONDON NEW YORK

First published in 1982 by
Chapman and Hall Ltd
11 New Fetter Lane
London EC4P 4EE
Published in the USA by
Chapman and Hall
733 Third Avenue
New York NY 10017

Printed in Great Britain by
J.W. Arrowsmith Ltd, Bristol

ISBN 0 412 22240 X

British Library Cataloguing in Publication Data
Moss, Robert
Animal population dynamics.—(Outline Studies in ecology)
1. Animal populations
I. Title II. Watson, Adam
III. Ollason, J. IV. Series
591.52′48 QH352

ISBN 0-412-22240-X

Library of Congress Cataloging in Publication Data
Moss, Robert
Animal population dynamics.
(Outline studies in ecology)
Bibliography: p.
Includes index.
1. Animal populations. I. Watson, Adam, 1930– .
II. Ollason, John. III. Title. IV. Series.
QL776.M67 1982 591.5′248 82–9531
ISBN 0-412-22240-X (pbk.)

Contents

Preface

Students of ecology set out to explain variations in the abundance of plants and animals. Population dynamics is an approach in which we concentrate on one or a few populations in a defined area. An understanding of what determines the size of such populations should help us understand what determines the size and distribution of all populations – that is, the pattern of living organisms on Earth.

The easiest way to teach a subject such as ecology is to present a theory, and then illustrate the theory with practical examples. From the point of view of tackling problems, however, a theory is simply a good way of organizing and predicting observations. Several alternative theories, each with quite different underlying assumptions, may fulfil this role equally well and be equally good at helping the student to learn a subject. A dialogue between theory and observation may eventually lead to the rejection of some theories and the confirmation of others, until these too are in their turn rejected.

This book is meant to be read by students and their teachers and to stimulate discussion between them. Concepts and theories are developed throughout the book, rather than presented as definitions illustrated with examples. The emphasis is on showing how concepts can be used to explain observations; and not on how the real world illustrates a particular set of ideas, nor on the people who developed these ideas. This approach also shows how concepts evolve and how those which are useful in one situation can be discarded in another.

Counts are the raw stuff of population dynamics and useful theories explaining variations in animal abundance can only be developed when proper measures of abundance have been made. In Chapter 1 we outline ways in which animals can be counted. We can view data on abundance by asking: how did the present patterns of abundance evolve? Chapter 2 sketches evolutionary aspects. We can also ask: what causes observed changes in abundance? The rest of the book is devoted to the proximate mechanisms causing changes in animal abundance.

At several points in the book, we end a discussion with an open question. At the moment, there is no one 'right' answer to most of these questions, and we hope that critical discussion between teachers and students will begin where we have left off, and lead to further study.

1 Counting methods

To study changes in numbers of animals, we must be able to count them, or at least obtain an index of their abundance. An early decision to make in most studies is what to count. The unit of study is often called a 'population' and natural boundaries to discrete populations of a convenient size for study do exist. Ants, seabirds and ground squirrels, for example, often live colonially and colonies are the obvious units to study. When studying animals with a continuous distribution over large areas, a worker is often forced to draw arbitrary limits on a map and to study the population inside these limits. Even this may be impracticable, as in the case of migratory marine fish: the only reasonable unit for studying the North Sea population of herrings *Clupea harengus* is the North Sea. And a colony of terns may inconvenience us by nesting on one shell spit in one year and moving miles down the coast to another in the next.

Despite these problems, we can count or estimate the number of animals present within an area at a given moment. Dividing the number by the area gives us the average density, a value which may be as or more important than the absolute size of the population.

1.1 Total counts

It is possible for an experienced worker to walk over an area of heather *Calluna vulgaris* moor with trained dogs, flush all the red grouse *Lagopus lagopus scoticus* there and count them as they fly off downwind (Fig. 1.1). The sea birds which gather in colonies to breed may be counted during the breeding season, as may seals. Barnacles attached to rocks may be counted when the tide is out.

Such easily-counted animals are a small proportion of the species studied. Even when a total count of a population is possible, it may require such effort that it is not worthwhile. Sampling may provide an estimate sufficiently accurate for the purposes of the study. Indeed, sampling a large area may be preferable to doing a total count on a small area. A total count may be useless if the area counted is too small to be representative of the population of interest.

1.2 Use of marked animals

It is often possible to capture, mark and release unharmed animals which cannot be counted directly. Bird leg rings, mammal ear tags and fish fin tags are widely used. In time, many of the animals in an

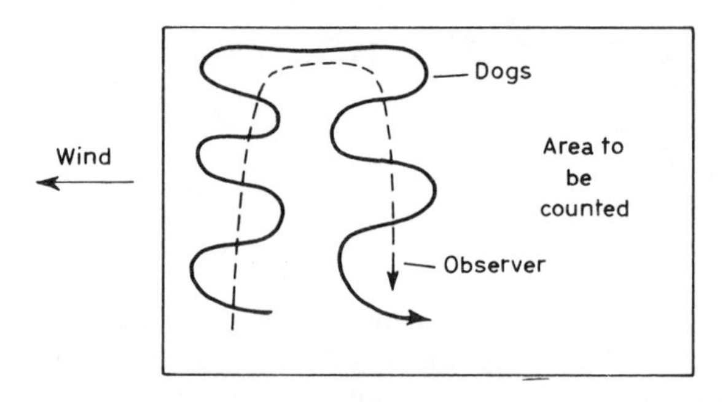

Fig. 1.1 Counting red grouse with dogs. The dogs flush the birds and the observer follows them with binoculars to ensure that they fly downwind and off the study area. If a bird flies upwind and lands on the as yet uncounted part of the area, this is noted and subtracted from the total number flushed. (Method described by Jenkins, Watson and Miller [1].)

intensively-trapped population will carry marks and so many of the animals captured will already be marked. It is then reasonable to assume that the proportion of new, unmarked animals caught in the latest trapping session is the same as that in the population as a whole. Knowing the number of animals marked, one can then easily estimate the total population. Say that we have marked a total of 50 animals and that we have just caught a sample of 20, 10 of which are marked. The total population is then $20/10 \times 50 = 100$.

Real studies are less simple than this. We have to be sure that all the marks remain on the animals. Furthermore, the population may not be static. Animals move or are born into a population, and emigrate or die. These gains and losses may be allowed for by appropriate corrections; indeed, data from an intensive capture/recapture programme can be used to estimate gains and losses [2].

A more subtle problem is the assumption which underlies such calculations, that marked and unmarked ânimals are equivalent in catchability and subsequent emigration or mortality. Strandgaard [3] made estimates of the number of roe deer *Capreolus capreolus* in a wood in Denmark; this was during the winter when the population was more or less stationary. He could watch the deer more easily than he could catch them and so he estimated the proportion of marked animals in the population from observations rather than recaptures. Young males were easier to catch than other deer, and young of both sexes were easier to observe than adults. Because of these and other biasses a fairly high proportion of the population had to be marked before a reliable estimate of the population was attained. Fig. 1.2 shows how the estimated number changed with an increasing number of marked animals. Strandgaard concluded that about two thirds of the population had to be marked before the estimated number was accurate enough for his purposes.

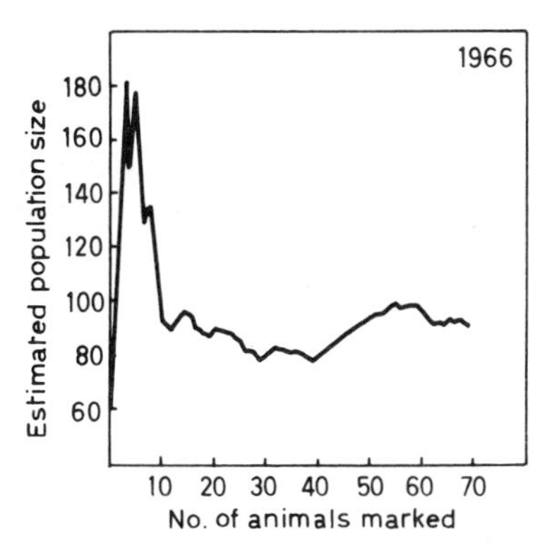

Fig. 1.2 The estimated population of roe deer at Kalø in 1966, calculated from an increasingly large marked proportion of the population. (After Strandgaard [3].) The best estimate of the population size, from a total enumeration of known animals, was 94.

1.3 Sampling by quadrats

A quadrat is a minute fraction of the total study area, so small that it is relatively easy to make a total count of animals within it. Such areas are often defined by means of a portable frame, also called a quadrat. The animals within a number of quadrats are counted and the mean density of animals within the quadrats is taken as the mean density within the study area.

If the quadrats have been placed randomly and if the animals have not run, swum, crawled or flown away whilst we were busy putting down the quadrat frame, then we can estimate the population density accurately, that is in an unbiased fashion. The more quadrats we measure, the more precise, that is repeatable, will our estimate be. More precision involves more effort – how much precision we need will depend on the question we are asking.

A major advantage of the quadrat method is that it allows the use of destructive sampling methods. Thus animals which live in the soil or mud and which cannot be counted directly can be studied; cores may be removed and taken back to the laboratory for detailed inspection. The assumption here is that removing a minute fraction of the population (and in some cases the habitat) does not affect the population in any important way. Problems in estimating abundance by quadrats may arise when the population is not distributed at random; animals may be clumped, for example, giving rise to a small number of quadrats with high counts and a large number with low or zero counts. Appropriate measures of abundance may be made by use of the appropriate statistical techniques [4].

The quadrat method can also be used to cut down on the time and effort required for a total count. Puffins *Fratercula arctica* are notoriously erratic in their attendance at colonies, and the number of birds on land or at sea near the colonies varies greatly both within a single day and between days. Harris and Murray [5] therefore decided to count occupied burrows, using quadrats rather than attempting the daunting and time-wasting task of counting thousands of burrows on precipitous terrain. Burrows, of course, are not birds but animal signs often provide indices of abundance where counts would be difficult or impossible.

1.4 Indices of abundance

All animals leave signs of their presence. Footprints, toothmarks, droppings, nests, casts and holes can all be counted to give some idea of variations in distribution and abundance. The relation between the number of such traces and the number of animals may be complex or unknown, and when this is the case such indices are of relative rather than absolute density.

A total count is usually the number of animals present at one time. Signs such as droppings accumulate over a period of time and can thus be used, for example, to indicate the relative times spent by a grazing animal on different parts of its home range. Betts [6] estimated numbers of canopy-feeding caterpillars (mainly of winter moth *Operophtera brumata*) by sampling the amount of their excreta falling onto trays beneath trees, and measuring their average rate of defaecation. Such techniques are particularly valuable for animals which feed at night and for those which are sparse or wary and therefore difficult to observe. Lockie [7] studied the ecology of pine martens *Martes martes* by signs alone, without seeing a marten for much of his study.

An index of abundance may in some ways be of more importance than the abundance itself. Marine laboratories throughout the world collect data from fishing vessels and express the abundance of a fish or whale stock as the catch per unit effort. For whales (Fig. 1.3) this is the number of whales caught per catcher-ton-day: the total catch divided by the number of days operating by catcher vessels (approximately the product of the number of catchers and the length of the season) multiplied by the average tonnage of the vessels. The tonnage is included as a correction for the increasing efficiency of the vessels.

Population biologists who study animals which vary greatly in abundance often use catch per unit effort as a method of measuring variations in animal density. This may be the number of insects caught per sweep of a net, the number of voles caught per trap per trapping session or the number of plankton per km of plankton net drag.

Whether a worker chooses to do total counts, to use a sampling technique, or to obtain an index of abundance will depend partly on the questions he wishes to answer, and partly on what is practically possible. The academic researcher interested in principles of population dynamics has often carefully chosen to study animals which are easy to count and relatively sedentary. He needs precise counts if he is to study detailed processes of population change. At the other extreme, those concerned for the future of the blue whale *Sibbaldus musculus* did not need to know

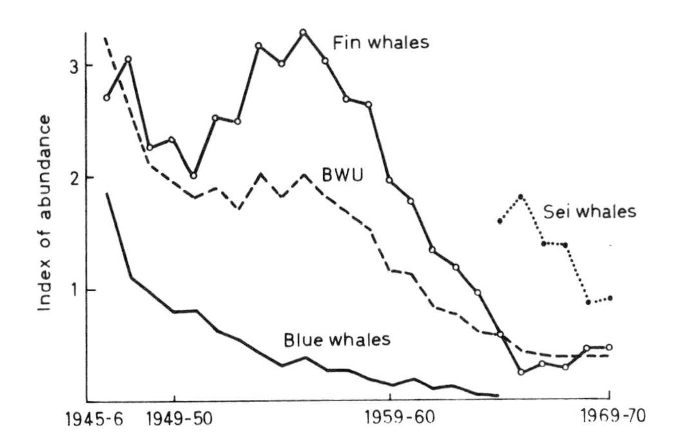

Fig. 1.3 Indices of abundance of baleen whales in the Antarctic, as estimated by the number caught in 1000 catcher-ton-days. BWU means Blue Whale Units, an index of the total catch. The apparent increase in fin whales up to 1956, and in sei whales up to 1966, is an artefact due to expeditions shifting their attentions from blue to fin and from fin to sei whales respectively. (After Gulland [8].)

whether the remaining stock was tens, hundreds or thousands. It was more important to show that the stock had been declining rapidly and was heading for extinction unless whaling could be controlled effectively (Fig. 1.3). The blue whale is now classified as 'protection stock' by the International Whaling Commission.

2 Evolution and population dynamics

Having made our counts, the next question we need to answer is: what determines the size and density of the population that we have counted?

2.1 What determines population size?

There are two completely different kinds of answer to this question. The two answers are really to the two different questions: why are populations of the observed size? and, what determines the size of observed populations? Answers to the first question are usually couched in evolutionary terms and comprise part of the subject of evolutionary ecology. The answer to the second question is birth, death and movement and is the main topic of this book. Of course, these two approaches are in no way competitive, rather complementary, and each should illuminate the other. In this chapter we sketch out some of the ideas about evolution which it is useful to bear in mind when thinking about population dynamics.

2.2 Natural selection

If we begin by assuming that the present distribution, densities and life-histories of living organisms have evolved by natural selection, we can ask what are the evolutionary pressures that have shaped the particular situation that we are studying. A simple question of this sort is: why are predators less common than their prey? The answer seems self-evident to the untrained mind. Predators must leave some of their prey uneaten, or the prey and then the predator would die out. Satisfaction with such an answer is an intellectual misdemeanour called teleology (the doctrine that nature shows evidence of design or purpose). The answer seems to imply that predators avoid increasing in abundance with the purpose of maintaining a healthy prey population. We can, however, re-phrase our answer more in accord with modern biological thinking.

A common starting point in such discussions is the doctrine of individual selection. An individual is regarded as a set of traits, such as body size, the tendency to produce a certain number of eggs, to be more or less aggressive or gregarious, to prefer certain habitats, to develop certain plumages and so on. Each trait is determined by the individual's genetic constitution and each individual is slightly different from all others. In a particular set of circumstances some individuals will survive and reproduce better than others, because the traits which they embody are better fitted to the environment. Hence these traits will become more common in the population as a whole; and this is the same as saying that the genes which are the bases of these traits are being selected for. From this point of view the individual and his relatives are important only as representatives of a set of genes, and 'individual selection' is seen as a battle for survival between competing genes.

How does this apply to predators and their prey? A predator which simply ate and reproduced would soon have so many descendants that they would all be likely to starve. A competing predator which took a territory and ejected its young when they were fully grown would be less likely to starve; and his offspring, inheriting similar traits, would also survive. Within limits, the bigger the territory, the better the chances of survival. Hence a genetic constitution which causes large territories would be selected for, and this would keep the predator sparse in relation to its prey in this case. A large number of such rationalizations is possible and students of evolutionary ecology are busy discussing and testing them.

2.3 The evolution of population characteristics

Different species of animal have evolved characteristic, and sometimes very different, life-histories. Evolutionary theorists generally start by assuming that almost every aspect of an animal's natural history, behaviour and physiology has a 'functional significance': that is, they are what they are because they help the collection of genes which comprise the animal to survive. The peculiarities of a given animal are thought to fit in to its life-cycle as a key fits a lock.

An important aspect of an animal's life-history is to survive competition from other species, competition which may be manifest as predation, disease, habitat destruction, parasitism, the production of toxins or antibodies, or competition for resources such as space, food and shelter. No two species have identical life-histories, and each is said to inhabit its own particular 'niche'. This word is usually thought of as the limits of the environmental variables that allow the species to exist and reproduce. Inhabiting different niches helps to avoid competition from other species; equally, competition from other species may force an animal to occupy a 'narrower' niche. An animal survives within its niche because it outperforms all competitors there. To exploit a particular niche effectively, animals have certain food preferences, body sizes, shapes, colours, behaviour patterns and appendages. Some animals are specialists, having very narrowly-defined requirements for survival, and others are generalists which have a 'wide' ecological niche.

An individual, however, may spend more time and effort competing with others of the same species than in surviving competition from other species. Many aspects of an animal's life-history may best be explained as resulting from the evolution of characteristics which endow the individuals possessing them with competitive advantages.

Important from our point of view is that species have evolved different and characteristic population densities (some always rare, some often abundant), longevities, reproductive rates and tendencies to disperse (emigration and immigration). These are the raw stuff of population dynamics. A useful start is to contrast '*r*-selected' and '*K*-selected' species. This dichotomy contrasts two ends of a spectrum, at one end of which are species with a high reproductive rate and short life-span (*r*-selected) and at the other end those having a low reproductive rate and long life-span (*K*-selected). The letters *r* and *K* refer to the logistic growth equation (Section 5.2). In general, *r*-selected animals can increase rapidly in numbers from a low density and are therefore well-adapted to patchy or rapidly-changing environments; they produce large numbers of offspring but put little effort into rearing each one. *K*-selected animals produce few offspring, but put much effort into rearing them; each individual is adapted to compete effectively for available resources. In a relatively stable environment a *K*-selected animal is likely to out-compete an *r*-selected one; but because of its low reproductive rate it is less able to take advantage of, for example, sudden flushes of food.

Hence different life-histories have evolved for explicable reasons. They also have important consequences for the general pattern of changes in animal numbers. Thus, the numbers of an *r*-selected animal will be likely to 'track' changes in the environment; that is, to change rapidly following a change in the environment. For example, if food is short, voles, with an average life-span of only a few weeks, will rapidly decline in numbers as breeding fails and adults die or emigrate. On the other hand, the *K*-selected owls which feed on the voles may not breed in years of food shortage, but will be able to survive until food becomes abundant again [9]. Hence the density of the adult population of owls may change little from year to year despite changes in the environment; if, however, food is insufficient even to maintain the adult population, then the *K*-selected owl will decline to low numbers and may take years to return to its original density.

2.4 The population as a unit for natural selection

There has been a prolonged controversy about the relative merits of individual as opposed to 'group' selection as bases for evolution by natural selection. Only one aspect of this need concern us here. The most highly-developed arguments for group selection depend on the 'structured deme' model for their exposition [10]. This is easy to envisage in a colonially-breeding species. Each colony has its own characteristic gene frequencies and few individuals move between colonies. Selection for one colony at the expense of another might occur if, for example, they were competing for the same food resource. Hence the genes in the successful colony might be selected for.

It is now becoming apparent that even animals with a fairly continuous distribution over large areas do not form a randomly-mixing population. Instead, there seems to be a mosaic, each piece of which is a group of animals more closely-related to each other than to other groups. We can imagine this as a number of colonies separated not by space but by more intangible barriers. Such notional colonies are the demes of the structured deme model, and the model suggests that selection can occur between them.

This brings us back to the problem of selecting a unit for study. In time it might become possible to identify local populations or demes which would then form a natural unit for the study of population dynamics. It seems likely that natural geographical features will tend to form boundaries between demes and such features are often used by biologists when defining a study area. This is something to bear in mind when looking for a study area.

Theories about the evolution of different life-histories are useful both as a framework within which to classify different species, and as a source of new ideas for testing. But they make little direct contribution to understanding the immediate causes of the particular changes in animal abundance that we see about us all the time. This is the subject-matter of the rest of this book.

3 The numerical analysis of population change

3.1 Correlations

In a population of constant size, with equal emigration and immigration, the birth rate must equal the death rate. If emigration and immigration remain equal and the population changes in size, the change must be equal to the difference between births and deaths. How do we determine the cause of an observed change in population? In theory it should be possible to do this without measuring births and deaths, but simply by recording the changing size of the population, or some index of it, and variations in the intensity of various possible causes.

Andersen [11] compiled records of the number of brown hares *Lepus capensis* shot on 22 estates in Denmark over a period of fifty years (Fig. 3.1). First, he noted a clear trend for the bag to increase over the period

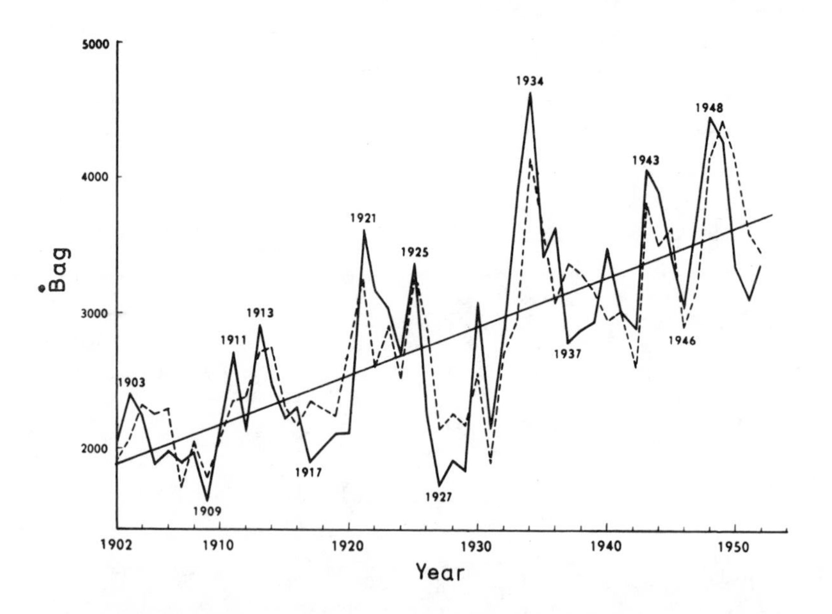

Fig. 3.1 Observed bag of hares at 22 estates in Denmark (full line), trend, and model (dashed line) retrodicting hare bags from the observed trend and a multiple regression of bags upon (i) rainfall in June–July, (ii) mean temperature March–June, (iii) frost-days December–March, (iv) autoregression (i.e. the bag in the previous year). (After Andersen [11].)

of the study. Next, he inspected weather records to see if these could explain the marked fluctuations about the trend. Bags tended to be high following warm winters (December–March) and springs (March–June), and summers (June and July) with low rainfall. Conversely, they were low after cold winters and springs and wet summers. There was also a slight tendency for bags to be high when the previous year's bag had also been high, and low when it had been low. Andersen constructed a multiple regression model which retrodicted (a new word meaning to predict after the event) deviations in hare bags from the trend by means of summer rainfall, the number of days with frost in winter, the mean temperature in spring and the bag the previous year. The model accounted for just over half the observed variance in hare bags. Evidently, much of the observed variation in numbers could be explained by variations in the rainfall and temperature.

The observed fit of the model with the observations was good and in this sense it was satisfactory. However, there are drawbacks to this approach. First, there is the technical problem that if we measure enough variables, then we are bound to come across some correlations sooner or later by pure chance and without any cause and effect relationship. Andersen spent years poring over weather records to find those combinations of data which best fitted the hare curve. This difficulty is even worse today when we have computers to do our poring for us. The best way round this is to build the model and then use it to make predictions of future changes in numbers. If the predictions work, this confirms the model and we can continue to use it with confidence.

The correlation between high bags of hares and dry summers was reasonable because hares breed in the summer and rain may well have killed some leverets. Similarly, cold winters and springs may have increased the mortality of fully-grown animals. However, the model does not test these suggestions directly, suggestions which are obviously important for a proper understanding of hare population dynamics. If we are going to study the effects of variables such as weather on breeding and mortality, we must have some direct measures of them. Furthermore, the chances of an animal breeding or dying are likely to vary with its age. Therefore we begin to take an interest in demography.

3.2 Demography

Birth, death and dispersal affect both the size and the age structure of a population, and the study of these processes is 'demography'.

3.2.1 Recruits

So far we have referred to birth rate, or 'natality', as a cause of population increase. However, man is usually interested in just one phase of an animal's life cycle. He wants to know what determines the number of fish over a certain size, the number of caterpillars eating his cabbages or the number of game birds available for shooting in the autumn. The word 'recruit' is used in such cases to mean a young individual or immigrant which enters the part of the population which is of interest, and which may not be the entire population. We may ignore the larval stages of a fish, for example, and only regard as recruits those individuals which reach a certain size or age class. Animals are recruited into the studied part of a population at a definite stage in their lives, but may die at any stage.

3.2.2 Survivorship curves

'The butterfly lives not months but moments, and has time enough' [12]. A field vole *Microtus* sp. can expect to live for a matter of weeks, whilst many fulmars *Fulmarus glacialis* and men live for decades. One limit to the length of time that an organism lives is its physiological lifespan. Laboratory mice provided with sufficient food and warmth will live for a year or two before senescence and death. Western man may well end his years in a geriatric ward after seventy years or more of active life. Such a pattern of mortality is illustrated by survivorship curve I in Fig. 3.2. A survivorship curve is simply a graph of the number of animals, present in the population at a chosen starting moment, and still alive at set intervals until there are none left. It is usual to begin with a cohort of 1000 young animals, newly recruited to the population of interest. Such curves are generally plotted using the logarithm of the number of animals; the slope of the curve is then a measure of the survival rate. Curve I shows a high survival rate for most of the lifespan followed by heavy mortality due to senescence.

Curve II, a straight line, shows a population with a constant rate of survival, and of course a constant mortality rate as mortality is simply the inverse of survival. Many bird populations show this kind of pattern, which implies that very few individuals ever reach their physiological life-span and that the chances of survival are independent of age. A simple situation resulting in a type II curve would be a population limited by competition for some constant resource such as breeding sites, in which each cohort of recruits is competing with preceding and

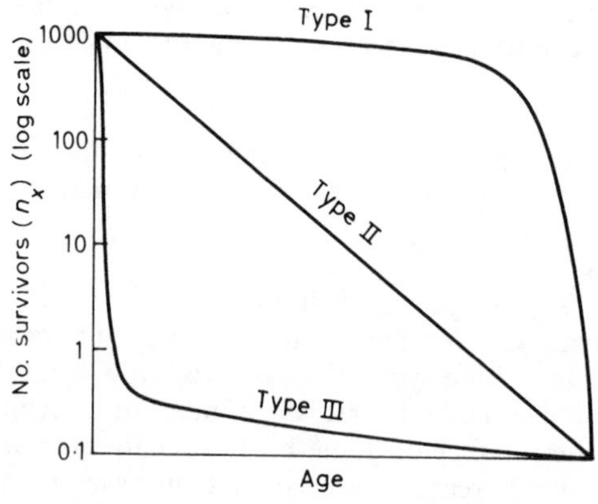

Fig. 3.2 Hypothetical survivorship curves. (After Pearl [13].)

succeeding cohorts on equal terms irrespective of age.

The third type of curve in Fig. 3.2 shows heavy mortality of young individuals followed by improved survival of older animals. Many marine fishes and invertebrates show survival curves of type III; small young animals are eaten and die, whilst bigger animals are less vulnerable to predation and better fitted to compete for survival.

In gross terms, each species of animal has a characteristic longevity and survivorship curve. Nonetheless, when we look at one species in more detail we are likely to find marked and informative differences amongst populations. Differences in three survivorship curves for red deer (Fig. 3.3) were related to the intensity of shooting by man [14]. When shooting pressure was light a type I curve was noted and mortality was attributed largely to senescence. When shooting was heavy a type II curve was seen, showing that many animals were killed before they would have died naturally. Old deer apparently died as their teeth became worn out and they were unable to chew coarse winter fodder effectively. These populations were studied in the absence of important natural predators. In the deer's primeval environment it is possible that wolves and bears would have taken the place of man as predator, but whether they would have had the same effect on the survivorship curves cannot be told from the data.

3.2.3 Life tables

With the data for constructing a survivorship curve, we also have all the

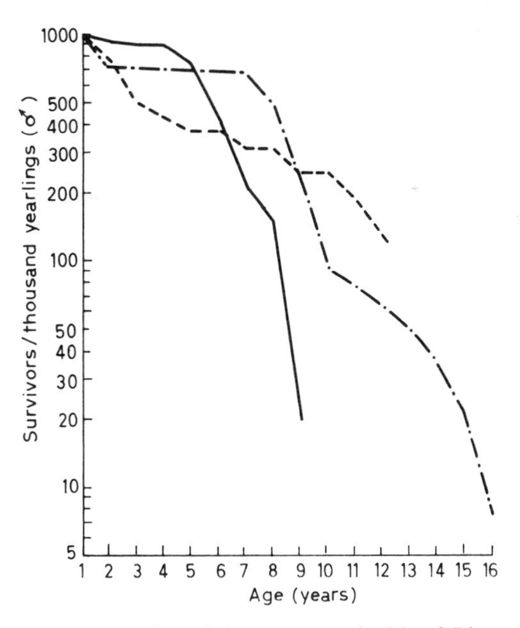

Fig. 3.3 Survivorship curves for red deer stags on the isle of Rhum (Scotland) and Germany. Stags on Rhum in 1957 (light shooting) —·—. Yearling stags on Rhum in 1957 (moderate shooting) ——. Stags in Germany (heavy shooting) - - - - -. (After Lowe [14].)

information needed to prepare a life table. The biggest users of life tables are insurance companies, who want to know how long the average human of a given age, sex and occupation is likely to live. Biologists are interested in similar questions about animal populations so that they can manage or understand them better. A life table summarizes the same survival data in different ways, ways which are designed to answer different questions about the population.

The first thing to decide is the age classes into which we are going to group the data. The class interval is often one year for animals which breed annually and live for several years, but may be a month or less for small rodents which can breed several times a year and may have a life expectancy measured in weeks. The symbols in general use are:

x = age interval
n_x = number alive at start of interval x
l_x = proportion alive at start of interval x, i.e. n_x/n
d_x = number dying between start of interval x and start of interval $x+1$, i.e. $n_x - n_{x+1}$
q_x = proportion dying during interval x, i.e. d_x/n_x
e_x = mean expectation of life for animals alive at start of interval x

The mean further expectation of life e_x is the parameter which comes nearest to answering the natural question about any animal: how long does it live? From values of the parameter n_x we can construct a survivorship curve, and also calculate all the other parameters in the life table. Similarly, even if we do not know n_x from direct observation, it and the entire life table can be constructed from knowledge of l_x, d_x or q_x. To standardize presentation, l_0 is often scaled up from 1.0 to 1000 (Table 3.1). Such a table is then for a notional cohort of $l_0 = 1000$ individuals and the n_x values are superfluous.

Table 3.1 Life table for the 1957 cohort of red deer stags on the isle of Rhum, Scotland. (After Lowe [14].) Values of n_x are not included and l_0 is set equal to 1000. Recruits were aged about 10 months, so the age interval $x=0$ was from 10 months to 1 year and 10 months and so on.

x *(age class)*	l_x *(survivors at start of age class x)*	d_x *(deaths)*	q_x *(mortality rate)*	e_x *(further expectation of life)*
0	1000	84	0.084	4.76
1	916	19	0.021	see
2	897	0	0	Table 3.2
3	897	150	0.167	
4	747	321	0.430	
5	426	218	0.512	
6	208	58	0.279	
7	150	130	0.867	
8	20	20	1.000	

To calculate e_x we have to do a few more sums. First work out the average number of individuals alive during each age interval; this is

$$L_x = (l_x + l_{x+1})/2$$

So from Table 3.1, L_0, the average number of stags alive in age class 0 to 1 years, is

$$L_0 = (1000 + 916)/2 = 958$$

Having calculated all the L_x we then work out T_x, the number of animal-years remaining to the cohort. This is done by adding up all the L_x between the x we are working on and the bottom of the table. In our example

$$T_5 = L_5 + L_6 + L_7 + L_8 = 591 \text{ deer-years}$$

Lastly, divide T_x by l_x to get e_x. *We have left Table 3.2 incomplete and suggest you work out the missing values.*

What we have just calculated is called a 'cohort' (or dynamic, age-specific, generation or horizontal) life table. The basic data were simply the number of red deer stag calves alive on Rhum in 1956 and still alive each year thereafter. To get this information Lowe had to wait until 1966, when they were almost all dead. Furthermore, he chose to work on an island so that the deer could not emigrate. Had he worked on the mainland, he would have been unable to follow the fate of individuals which moved out of the population.

An alternative method of deriving a life table is to estimate the age structure of the population at one moment. If our population has a stationary size over the years and also has constant and equal recruitment and mortality rates, then this 'static' (or stationary, time specific or vertical) life table should be identical with the cohort life table. In each successive year (or other time interval x), the number of animals of a given age would be the same. Instead of waiting for an entire cohort to die off, we could simply write in the observed number of each age class for each n_x and divide this by n_0 to get l_x. In theory, the fact that some individuals of a given age class have emigrated is

Table 3.2 Calculation of e_x from Table 3.1.

x	l_x	L_x	T_x	e_x
0	1000	958		4.76
1	916	907		
2	897	897		
3	897	822		
4	747	587	1178	1.58
5	426	317	591	1.40
6	208	179		
7	150	85		
8	20	10		

unimportant, as they should have been replaced by an equal number of immigrants of the same age.

Andersen [15] exterminated all the roe deer on Kalø estate in Denmark. When the kill was stopped, 213 roe deer had been shot in the 340 ha of woodland. These were aged according to the degree of wear on their molars and the results are shown in Fig. 3.4. If the assumptions in the last paragraph are made, then the data are equivalent to the dynamic life table which would have been obtained by following a cohort throughout its existence. *You might like to do this.*

It is not necessary to kill an entire population in order to construct a static life table. This can as well be done on the basis of a cross section of the population obtained by random sampling. Marine biologists assume that the fish caught by fishing vessels are a random sample of the population of interest and they construct life tables from a subsample of the commercial catch. Some fish scales, like trees, grow an extra ring each year and we can count these rings to age the fish.

In fact it is rare for cohort and static life tables to give identical results, because most populations are not stationary and because reproduction and mortality generally vary from year to year. Hence life tables derived from static age-distributions, as well as differing from cohort life tables, may vary from year to year and place to place. Used judiciously,

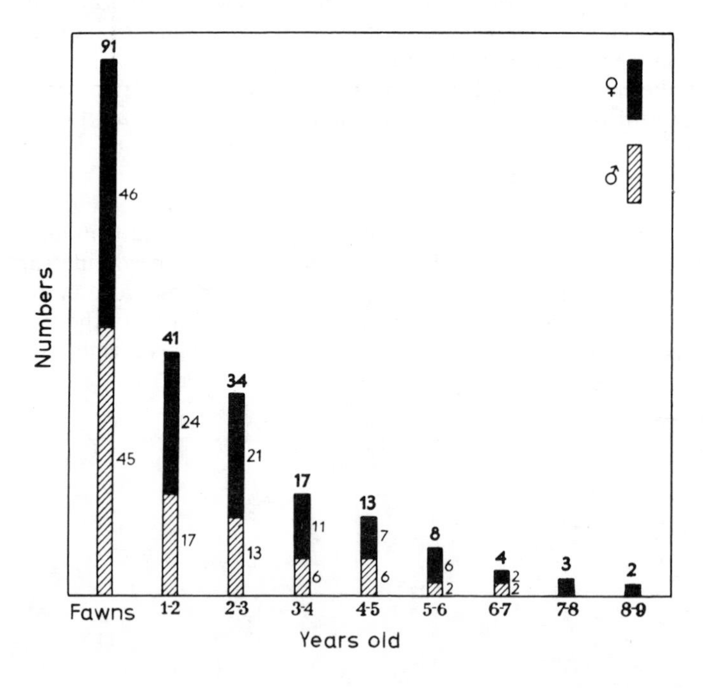

Fig. 3.4 Age and sex distribution in the total population of 213 roe deer killed at Kalø, Denmark, in 1950. (After Andersen [15].)

however, differences between static age-distributions can provide useful information about the dynamics of a population.

The catch of anchovies *Engraulis mordax* landed in California in 1973–4 was a record, due partly to the very 'strong' 1971 year-class (Fig. 3.5). This year-class had also predominated in 1972–3 and 1971–2. Such strong year-classes are typical of marine fish stocks, and understanding the causes of this is a major concern in marine biology (see Chapter 6).

When animals are rare or elusive, data may be too sparse to construct proper life tables of either the cohort or the static variety. Nonetheless, it may be possible to add several years' data together to get a general picture of a notional average or typical cohort. Romanov [17] did this to show that capercaillie *Tetrao urogallus* bred well in some parts of their range and poorly in others. In the Kirov region in the southern taiga, an average 69% of the autumn population comprised young birds hatched earlier in the year; this contrasted with the Ryazan region in the northern edges of the taiga where the average proportion of young was only 34%. Capercaillie numbers fluctuated more from year to year at Ryazan but showed no marked trend over the period of study at either area. On

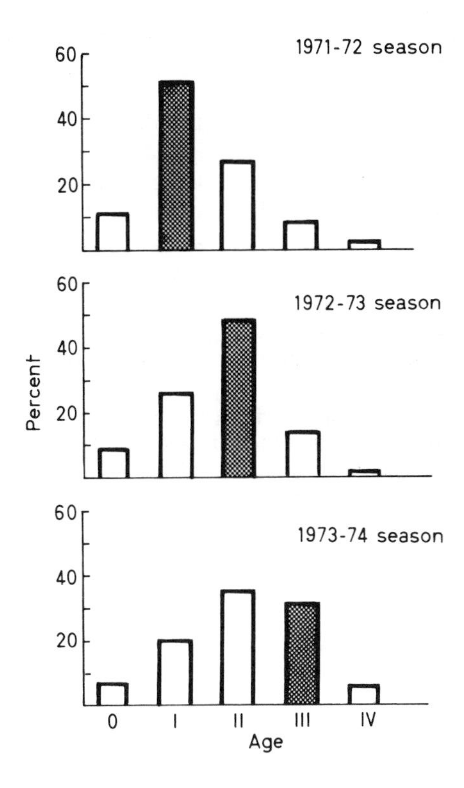

Fig. 3.5 Age class composition (by numbers) of anchovies landed at San Pedro, California. (After Sunada [16].)

average, therefore, recruitment to both autumn populations was equal to losses. Average recruitment to the 'good' and 'poor' populations was 69% and 34% respectively, and the same must have been true for losses. Furthermore, it is reasonable to assume for a game-bird such as the capercaillie that survivorship follows the pattern of curve type II (Fig. 3.2), that is, mortality is constant irrespective of age. Hence the life expectancy of capercaillie on the poor area, where population density was also relatively low, should be greater than that on the good area.

Romanov also marked hundreds of birds and recorded when he last saw them. In Table 3.3 we show the results of his ringing programme, laid out as if we were about to calculate a cohort life table, and contrast these results with those calculated from constant losses of 69% and 34%. Clearly the results from the two different kinds of data are quite different, a puzzling observation. Nonetheless, marked birds did live longer on the poor area than on the good area, as predicted from the age structure. This is the kind of information we are often presented with in the real world: partial confirmation of one hypothesis accompanied by data that can only be explained by further work in the field. Romanov's conclusions, based on this and other data, were as follows. Much emigration of young birds occurs from both areas, but this is exceeded by immigration into the poor area. Most birds which emigrate from the poor area do so in their first year, but birds hatched in the good area usually wait until their second year before leaving. Thus survivorship curves constructed from the columns marked (a) in Table 3.3 would not represent survival at all, but rather the proportion of birds in the marked segment of the population still alive within the studied area.

Can you offer alternative interpretations of the data? What information would you collect to distinguish between alternatives?

3.2.4 k-*factor analysis*

Life tables are useful descriptions of survival and mortality, but they are

Table 3.3 Survival of capercaillie within two populations, as shown (a) by observations on marked birds and (b) by the average age structure of the populations, assuming constant mortality. (Data from Romanov [17].)

Age class (years)	*Kirov region (sample of 210*		*Ryazan region (sample of 134)*	
	(a)	*(b)*	*(a)*	*(b)*
0	1000	1000	1000	1000
1	743	310	425	660
2	172	96	306	436
3	52	30	150	298
4	19	9	75	190
5	9	3	30	125
6	5	1	7	83
7	0	0	7	55

not well adapted to explaining causes of changes in density from one year to the next. Yet this is a problem of immediate interest to layman and biologist alike. Why are there so many wasps eating my apples this year, when there were so few last year? What has happened to all the swallows this year? Striking questions that are usually difficult to answer. When presented with a difficult question, the first job of the scientist should be to define exactly what it is that he has to explain. In k-factor analysis we tackle this task by breaking down gains and losses to the population into a number of parts and then analysing the relative importance of these different parts or factors to population change. ('k-factor' is shorthand for 'killing factor' and should not be confused with K, the 'carrying capacity' in the logistic equation (Section 5.2) or the term 'K-selected', despite the unfortunate similarity.)

Imagine a population of 1000 animals at the start of a year, animals which have clearly defined stages in their annual cycle. This might include eggs, several larval, and pupal and adult stages. It is reasonable to consider the losses to each successive stage separately. However, the loss of 100 eggs, say, will have a smaller impact on the eventual population than the loss of 100 adults; because many of the eggs would have died before they reached the adult stage, 100 adults are equivalent to more than 100 eggs. We can make losses during each stage comparable by expressing each loss as a proportion of the numbers present at the beginning of that stage. We can then multiply all the individual losses together to get the total loss, expressed as a proportion of the initial population. A great advantage of this approach is that it forces the worker to consider every part of the animal's life-cycle as a potential cause of loss. Without this discipline, it is tempting to ignore those stages or seasons for which we have poor data or no data.

To make subsequent manipulation easier, it is usual to express the counts in logarithms. This allows us to add and subtract, which is easier than multiplying and dividing. Losses are then worked out by subtracting the logarithm of each count from the logarithm of the previous one, and the total loss by adding up the individual losses. A worked example is given in Table 3.4. This procedure is repeated for each year's data and the results plotted as in Fig. 3.6 which shows a k-factor analysis for the winter moth studied by Varley and Gradwell [18]. The flightless females

Table 3.4 Calculations of k-factors from hypothetical counts

Stage	*Count*	*Log_{10} count*	k-*factor*
1	1000	3.00	
			0.30 (k_1)
2	500	2.70	
			0.05 (k_2)
3	450	2.65	
			0.06 (k_3)
4	390	2.59	
			0.01 (k_4)
5	380	2.58	
			0.68 (k_5)
6	80	1.90	
		$k_1+k_2+k_3+k_4+k_5 = 1.10$ (total K)	

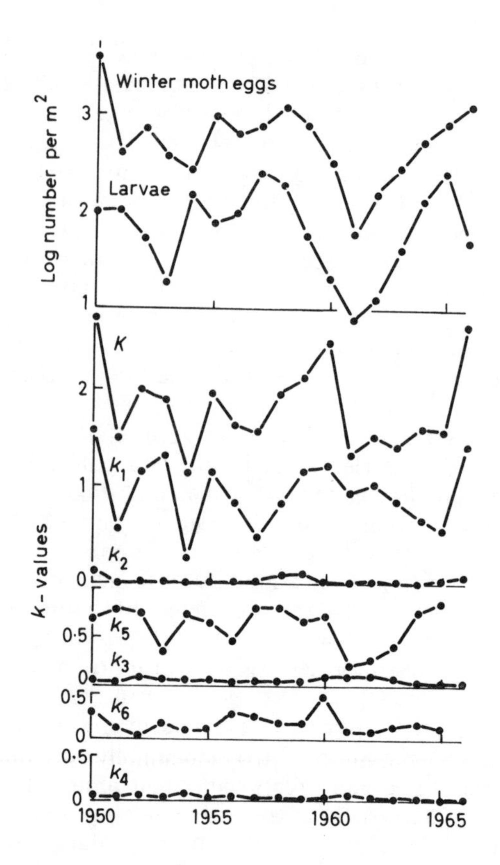

Fig. 3.6 Graphical k-factor analysis of population dynamics of the winter moth. (After Varley and Gradwell [18].)

climb the trunks of oak *Quercus robur* trees in November and December before laying eggs high in the canopy. The eggs hatch when the oak buds are bursting in spring; larvae crawl into the buds, feed on the leaves and are fully grown in May. They drop to the ground on silken threads and pupate in the earth where they remain until emergence as moths next winter. Varley and Gradwell identified six categories of mortality or k-factors:

k_1 'winter disappearance' including the mortality of adult moths before they had completed egg-laying in autumn, plus egg mortality and early larval mortality.

k_2, k_3 and k_4 were the effects of various parasites and parasitoids on the larvae. In fact these did not operate strictly in sequence as assumed by the model, but the data were treated as if they did.

k_5 was 'pupal predation' by beetles and shrews and k_6 the effect of an ichneumonid parasitoid. Again, these data were treated as if they acted sequentially.

The results (Fig. 3.6) are striking. When the total mortality K is reckoned from one egg stage to the next, one immediately sees that the greater part of the changes in K is caused by changes in k_1 or winter disappearance. Most of this is probably 'hatchling mortality' when the young larvae are looking for oak buds in which to establish themselves. The conclusion is clear: to understand the population dynamics of the winter moth, attention must first be directed to finding the causes of hatchling mortality.

A k-factor analysis is a pleasingly simple way of presenting masses of data which might otherwise overwhelm the investigator. It is particularly applicable to animals, such as the winter moth, with simple life histories and non-overlapping generations. However, even in the analysis above, certain unrealistic assumptions had to be made; and as life histories get more complex, so assumptions become more unreal and k-factor analysis becomes less applicable. It is rarely used for mammals, but, with caution [19], can be applied to some bird populations. Nonetheless, the general idea of mortalities acting sequentially at successive stages in an animal's life history is a valuable one which helps to organize our thoughts even when a simple k-factor analysis is not directly applicable.

3.3 Population regulation

If all the individuals in a population reproduced at their maximum possible rate, and all survived to their physiological lifespans, that population would continue to grow indefinitely. To find out what processes limit or regulate populations to their observed sizes, we first define and measure mortality factors. In k-factor analysis, these can simply be the losses occurring during each stage of an animal's life history; and failure to reproduce at the maximum possible rate is regarded as a potential loss to the population. Population ecologists sometimes slip into the habit of thinking of a reduced rate of breeding as just another mortality factor, and we shall sometimes use this convention.

Other means of classifying mortality may be used. For example, we may count the number of animals eaten by predators, the number dying from disease and the number starving from lack of food, and call 'predation', 'disease' and 'starvation' mortality factors. A mortality factor is usually defined in terms convenient to a particular study and is always arbitrary. Furthermore, mortality factors may not be independent of each other: if starvation kills some larvae, surviving poorly-fed individuals may metamorphose into small adults which lay few eggs. Conversely, mortality due to one factor may reduce mortality due to another: heavy predation may reduce a population so that fewer individuals starve when food is short. Such inverse relationships between mortality factors are termed 'compensatory' and we shall return to this point.

As well as setting upper limits to population size, mortality may cause the extinction of a population. In 1968 the total number of California condors *Gymnogyps californianus* was about 55 individuals and in 1978 about 30 [20]. The bird is long-lived and the population would require to fledge only about four chicks a year to maintain a size of about 50 birds. For the last 20 years, however, the average has been about 1.25 chicks. Mortality due to disturbance by humans, added to mortality occurring for natural reasons, has been enough to cause the decline of the condor.

A species can survive by two mechanisms, and both probably operate in nature:

3.3.1 Species which survive by dispersal

Local populations may well become extinct. A species with a reasonably wide distribution will have many local populations and catastrophes are unlikely to affect them all simultaneously. A cereal farmer may exterminate the aphids in a field, but immigrants will soon found a new population. Aphids, like many other *r*-selected insects, have a winged phase which is well-suited to dispersal by the wind. In general, many species are adapted to local extinctions and have a life-history which includes much emigration.

Some animals have a life-history which inevitably results in almost any local population being likely to become extinct in the near future: such species rely on transient centres of population as a source of migrants which found new colonies. After a hurricane, certain tropical corals rapidly colonize newly-bared areas of reef, grow quickly, and emit potential colonists in the form of free-swimming larvae. Other corals, slower to colonize but better at competing for space, gradually take over the habitat as the years pass. Eventually, the original colonists are excluded. Diversity on a tropical coral reef may well be maintained by recurrent catastrophes [21].

Such transient populations often depend on and occur in a particular stage in a seral succession. The black swallowtail butterfly *Papilio polyxenes* in the humid montane regions of Costa Rica [22] depends almost entirely on the umbelliferous plant *Spananthe paniculata*. This, in turn, germinates after any disturbance that creates bare soil and is common in abandoned fields. After about six months the plant dies and is replaced by the next stage in the seral succession. Both plant and butterfly must then find a new patch of ground on which to grow.

3.3.2 Survival of populations by regulation

The observed reproductive rate of many populations is below what is physiologically possible. In a pack of wolves only the dominant female may breed even though the subordinate females are physically mature [23]. When a natural population of fish, seals or whales is first exploited, the age at which individuals first breed decreases. Such a response tends to compensate for the additional mortality and is therefore 'compensatory'.

In a population of capercaillie which bred poorly, adults tended to live longer than in a population which bred well (Table 3.3). Another compensatory relationship was in a population of the spittle bug *Neophilaenus lineatus* on lowland grassland; in years when many nymphs died, a higher proportion than usual of the adults survived [24]. Conversely, when the nymphs survived well, more adults than usual died. The reasons for such inverse relationships are not always clear, but their effect is to help stabilize a population.

Observations of compensation in mortality and reproduction lead naturally to the idea that populations exhibit some kind of homeostasis; as does the simpler observation that many populations remain remarkably steady in density from one year to the next. Postulated homeostatic mechanisms are thought to 'regulate' animal densities. One school of thought, which many entomologists have found valuable, regards density dependence (see below) as the key to understanding regulation. In the rest of this section we give a simple version of this viewpoint. In Chapter 4, we give a different point of view.

If a population is self-contained, with no movement in or out, then compensatory reproduction and survival are necessary if it is to avoid extinction. After a catastrophe which has reduced such a population to a low level, the survivors must be able to breed faster or survive longer than average; population density then returns to its usual level. Conversely, when densities are higher than usual, reproduction may decline and mortality increase until density returns, once more, to its usual level. An example of these processes might be that some limiting resource such as food is in short supply at high densities. At low densities, each animal has more than enough and survives and produces many young; at high densities, competition for food results in animals starving or failing to breed.

In a constant environment such a population is likely to reach and remain at its usual or 'equilibrium' density. If some random process causes densities to depart from equilibrium, then the homeostatic, regulatory mechanisms will tend to return it. This suggests that we should be able to detect such regulatory processes by their tendency to return the population to equilibrium, and distinguish them from random, non-regulatory, processes which do not.

If a rate of mortality increases at high densities and decreases at low densities, it is said to be 'density-dependent'. Note that the rate is important and not the total mortality. If the proportion of the population which dies remains constant, then the number dying will increase with increasing density; but this is not density dependence (Fig. 3.7) and will have no regulatory effect on the population. In this context, the word 'regulation' has a specific meaning and applies to homeostatic mortality factors which operate in a density-dependent fashion and are effective, either alone or in conjunction with other density-dependent factors, in reducing the population from high densities and increasing it from low ones. Such a population is constantly tending to return to its equilibrium density and only fails to reach it because of random

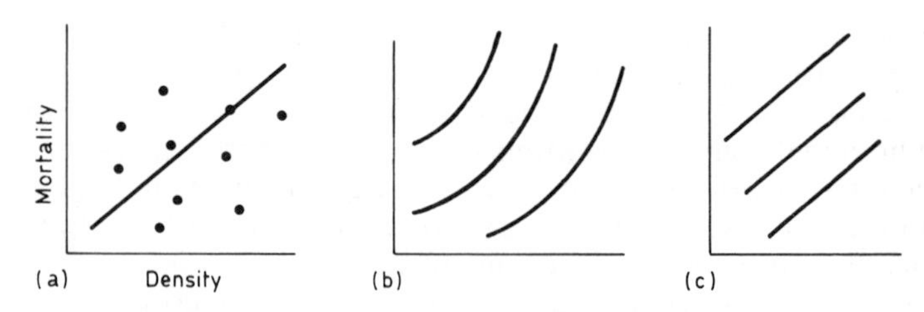

Fig. 3.7 Mortality (a) as a constant (line) or random (spots) proportion of density, i.e. density-independent; (b) mortality rates increasing with density, i.e. density-dependent; (c) is the same as (b) but on a logarithmic scale and shows that the numerical effect of a density-dependent mortality factor can be measured as the slope of a straight line on a logarithmic plot.

'density-independent' disturbances. 'Limiting' factors, on the other hand, are those which have limited the population in a given season and may or may not be regulatory [25].

We can clarify this by taking another look at data from the winter moth. From Fig. 3.6 it is clear that k_1, winter disappearance, is the main cause of variations in winter moth numbers. However, when k_1 is plotted against the logarithm of the density of the population (in this case, nearly enough, the population of newly-hatched larvae) it shows no density dependence (Fig. 3.8(a)). Therefore, in the terminology of the last paragraph, k_1 is the main limiting factor but not a regulatory one. When there is one limiting factor of major importance, like k_1 here, this is sometimes (somewhat confusingly) referred to as the 'key factor'.

The factor k_5, pupal predation, on the other hand does show clear density dependence when plotted against the logarithm of the initial density of pupae (Fig. 3.8(b)). Therefore k_5 may be a regulatory factor. Varley and Gradwell concluded that 'without the stabilizing action of the density-dependent factor the winter moth would not be regulated'. This of course implies that the population was indeed regulated and this was not established. Unquestionably, k_5 tended to reduce high densities, was less important at low densities and was able to regulate a

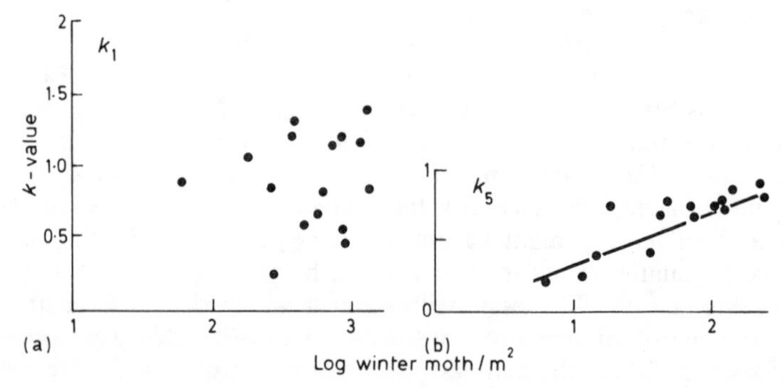

Fig. 3.8 Plots of (a) k_1 and (b) k_5 against the logarithm of the densities on which they act (winter moth, after Varley and Gradwell [18].)

mathematical model of the population. Nonetheless, this model involved some untested assumptions and for this reason it was not shown unequivocally that k_5 was sufficient to stabilize the population about an equilibrium density, towards which it would tend to return irrespective of density-independent factors such as k_1. It might have been fluctuating at random.

What would you have done to test whether the population was regulated or not? Would you be satisfied with purely numerical analyses or would you prefer to tackle the question experimentally?

Populations can be regulated by density-dependent mortality and this has often been shown in the laboratory. Also, there are many examples of density-dependent mortality factors operating in the field. However, it has proved extremely difficult to detect density-dependent regulation in natural populations. One reason for this is that a constant equilibrium density is unlikely in a natural situation because the habitat and environment change from year to year. Therefore, the same observed density may be above the varying equilibrium density in one year and below it in the next. In theory, one could get round this problem by plotting the strength of a mortality factor against the difference between the observed density and the equilibrium density. Unfortunately, the equilibrium density is a purely theoretical concept and cannot be measured in the field.

Another problem is that it is difficult to be sure that a density-dependent factor is also regulatory when the main limiting factor, as is often true in invertebrates, seems to be density-independent. Finally, ecologists usually classify different kinds of mortality in arbitrary categories which are easy to measure and then call these 'mortality factors'. Any such category, be it the mortality occurring during a stage in the life history of an arthropod, the mortality affecting a vertebrate during a particular season, or mortality due to disease or predation, may well have both density-dependent and density-independent components.

3.3.3 Inverse and delayed density dependence

The unqualified term density dependence is usually taken to mean direct density dependence and that is the sense in which we use it. There are other possible relationships between density and mortality factors. 'Inverse density dependence' occurs when a mortality factor operates more strongly at low densities than at high densities. Nymphal mortality in spittle bugs *Neophilaenus* and *Philaenus* spp. sometimes shows such a pattern; the reason might be that predators have a bigger impact on sparse populations than on dense ones, but this is not clear [24, 26]. Whatever the cause, models of population regulation suggest that inversely density-dependent mortality tends to destabilize a population, causing it to fall at low densities and rise at high densities unless other factors compensate.

'Delayed density dependence' is also observed in natural populations.

This occurs when there is a lag in the operation of a factor, so that it relates to the densities at some time in the past, rather than present ones. Predation on snowshoe hares *Lepus americanus* shows delayed density-dependent characteristics. Some populations of this hare show marked cycles in numbers with a period of about 10 years. As numbers build up, owls, hawks, foxes and lynxes *Lynx canadensis* gather to feed on the abundant prey. When the cycle starts on its downward swing, the ratio of prey to predators decreases and so the proportion of hares eaten by predators increases [27]. Thus predation pressure on the hare population is related to hare densities two years or so before the current year. Delayed density dependence of mortality factors is typical of cyclic or oscillating populations.

3.3.4 Populations at the centre and edge of the species' range

Despite the criticisms outlined in Section 3.3.2, students of arthropod populations usually accept that a strongly density-dependent mortality factor is likely to be regulatory. A simple test for detecting density-dependent mortality and possible regulation is based on the idea that density-independent mortality factors should act randomly, irrespective of the population density. Therefore, if we plot the logarithm of population density in one year ($\log P_t$) against that in the following year ($\log P_{t+1}$) the slope of the line best fitting these points should be 1.0. In a regulated population, however, the slope of the line would be less than 1.0 and the amount by which it falls short should be a measure of the degree of regulation [28].

Whittaker [24] studied the spittle bug *Neophilaenus lineatus* both in a lowland grassland in the centre of its range, and in an upland grazing at the edge of its altitudinal limit. He found that the edge population fluctuated more from year to year than did the central one; this is characteristic of such contrasts and similar to the capercaillie (Section 3.2.3). When Whittaker plotted spittle bug densities in one year against

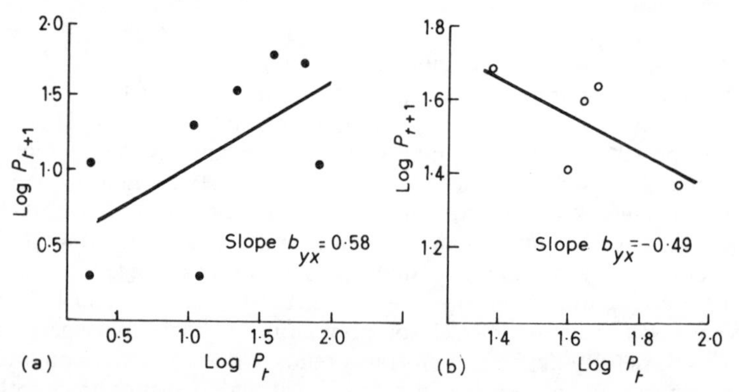

Fig. 3.9 The relation between numbers of spittle bugs in one generation ($\log P_t$) and those in the next ($\log P_{t+1}$). (a) In a population at the edge of its range and (b) in a central population. (After Whittaker [24].)

those in the next (Fig. 3.9), the slope for the edge population did not depart markedly from 1.0, but that for the central population was negative. This was evidence for density dependence, and further analysis led him to conclude that the central population was regulated by compensatory density-dependent adult mortality. An important component in this mortality was attack by the dipteran parasitoid *Verrallia aucta*. The edge population, on the other hand, showed little evidence of regulation and occurred in an area where local extinctions were common. At the edge of its range, the spittle bug maintained its presence by immigration.

To test the conclusion that spittle bugs in the edge population were not regulated, Whittaker removed half the population (Table 3.5). He did this in a period of natural increase in numbers, as shown by the control, but density on the experimental area failed to increase. The experiment had confirmed the conclusion.

Unfortunately, Whittaker was unable to do a similar experiment on the central population. *What would the result of such an experiment have been if indeed the population was regulated?*

The study on spittle bugs illustrates an important general principle. In the centre of their range, in habitats to which they are well adapted, animals can produce many offspring but increases in numbers are generally small at average densities and very small at high densities. When some mortality factor increases above its usual intensity, another mortality factor decreases to compensate for it. So the population persists and often remains remarkably steady in numbers for many years. In poorer habitat at the edge of its range, many mortality factors act more intensely than at the centre. Mortality may exceed the population's capacity to compensate, and extinctions are likely. Thus the edge of a species' range occurs where the frequency of extinctions exceeds the rate at which immigrants found new populations.

There has been much argument between people who have studied central and those who have studied edge populations. Students of central populations emphasized the importance of regulation in the control of animal numbers, whilst students of edge and transient populations were more interested in the balance between those factors which allowed an animal to reproduce and survive and those which

Table 3.5 Changes in the density (no. of instar 2 larvae/m^2) of an artificially reduced and a control population of spittle bugs at the edge of their range. (After Whittaker, 1971 [24].)

Year	*Experiment*	*Control*
1965	16	11
1966	29	21
1967	37	37
Removal of larvae from the experimental population		
1968	18	65
1969	25	56

killed it. The work on spittle bugs allows us to synthesize these viewpoints into a more comprehensive understanding.

There is a remaining conflict, however. If regulation does occur, but in relation to a varying equilibrium density, then the attempt to detect regulatory factors by looking for density dependence may be of little value. This is because the mean density observed during a study may bear little relation to the varying equilibrium. Students of mammal populations rarely emphasize density dependence. If, however, fluctuations in density are so large that variations in equilibrium density are unimportant in relation to observed changes in numbers, then the detection of density-dependent factors may well be useful for studying regulation. Entomologists often find this to be the case.

4 The natural limitation of animal numbers

A task of the population biologist is to assess the relative importance of weather, food, competitors and a place to live in determining the abundance of the animals he is studying. One approach, outlined in Chapter 3, is to distinguish between regulatory (density-dependent) and random (density-independent) mortality factors. It is assumed that random factors can cause population density to move away from an hypothetical equilibrium density, whilst regulatory factors return it.

This approach often works with invertebrates but is seldom used with vertebrates. An alternative is to ask the empirical question: what factors have limited population density during a particular study? From many such observations we may be able to infer some general principles. In particular, we may be able to say what limits animal abundance without the need to demonstrate formal density dependence.

4.1 Limiting resources

4.1.1 Food

In 1944 the United States Coast Guard introduced 27 reindeer *Rangifer tarandus* to St Matthew Island in the Bering Sea [29]. By 1963 these had increased to 6000 (Fig. 4.1); in 1966 only 42 remained after a massive die-

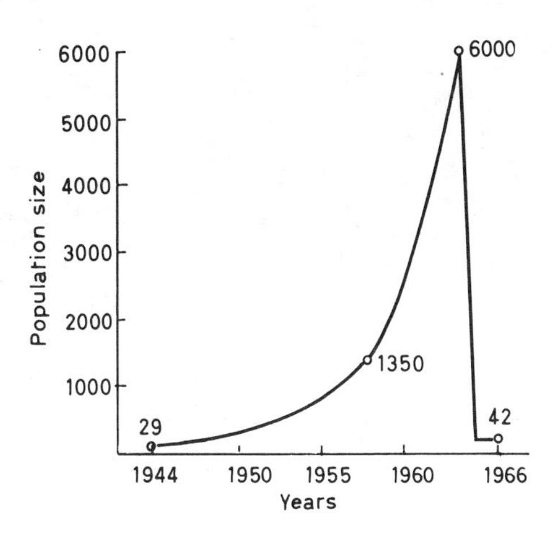

Fig. 4.1 Assumed population growth of the St Matthew Island reindeer herd. Actual counts are indicated on the population curve. (After Klein [29].)

off in the winter of 1963–4. Lichen, usually the most important winter forage of reindeer, was reduced from a mat about 10 cm deep to badly shattered fragments. No fat was present in the marrow of the reindeer's long bones, showing that they had starved. Overgrazing, assisted by deep snow in the winter of 1963–4, was the cause of this catastrophe.

Many winter die-offs of deer have been reported, usually with deep snow and high population densities as predisposing causes. In the winter of 1948–9 over a third of the Jawbone herd of mule deer *Odocoileus hemionus* in California perished [30]. Many individuals became too weak to wade through the heavy snow and could be run down and caught by hand. After hundreds of fawns had died the does deserted their normal ridge-top ranges to seek food and shelter in nearby canyons. Many of the refugees were too weak to climb uphill and moved down to the river banks, where their carcasses accumulated. *Post mortem* examinations showed no sign of epidemic disease and no cause of death other than starvation. Buck brush *Ceanothus cuneatus*, the deer's preferred food, was heavily browsed. Some plants were severely hedged and died and the observers found very few seedlings during their three year study. The productivity of surviving buck brush plants decreased as they became weakened by heavy browsing (Fig. 4.2).

Recurrent catastrophic starvation is a feature of the cinnabar moth *Tyria jacobaeae*. Some populations build up until they completely defoliate their food plant, ragwort *Senecio jacobea*. Many larvae starve

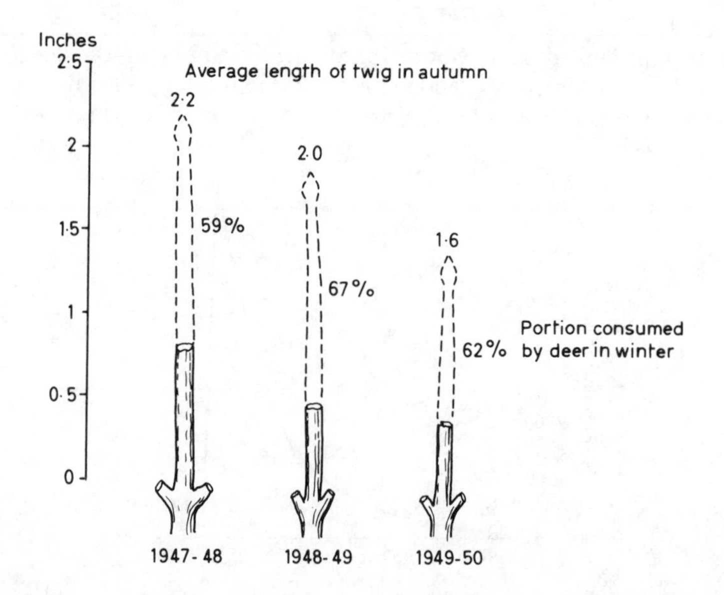

Fig. 4.2 Annual growth of buck brush twigs, and the percentage browsed by mule deer of the Jawbone herd, California, in three years. (After Leopold, Riney, McCain and Tevis [30].)

and the few survivors metamorphose into adults which lay few eggs. This gives the ragwort a chance to recover until the moth again builds up to high density. The resulting fluctuation is modified by the growth of ragwort also being dependent upon rainfall. Dempster and Lakhani [31] accounted for most of the observed variation in cinnabar moth numbers in a model which included rainfall and the effect of food supplies on moth recruitment and losses. Starvation alone was sufficient to give a quantitative account of the dynamics of this population of the cinnabar moth.

In the examples above, reindeer, mule deer and cinnabar moth were all clearly limited in density by starvation. This was obvious because they ate most of the food available to them and then died in large numbers with clear physical signs of malnutrition. Many animals, on the other hand, do not eat all the food available. Wolves maintain a fairly steady population in the presence of abundant prey, and remain in good condition [23]. Birds of the grouse family (*Tetraonidae*) eat a minute fraction of the forage available to them, often less than 5% [32]. We have to invoke more than food shortage to account for the many such observations.

4.1.2 Limiting factors other than food

Weather can be important, either by itself or through its effects on food supplies. A hurricane can destroy entire populations of coral on a reef. Icy snow can seal off abundant food underneath, or accentuate the effects of food shortage, as in the reindeer and mule deer above. Heavy mortality of ruffed grouse *Bonasa umbellus* occurred when freezing rain covered the tree buds that form their winter food with a layer of ice [33]. When rains fall in the Arizona desert, green plants grow and the Gambel quail *Lophortyx gambelii* breeds well; when the rains fail the quail breed badly (Fig. 4.3).

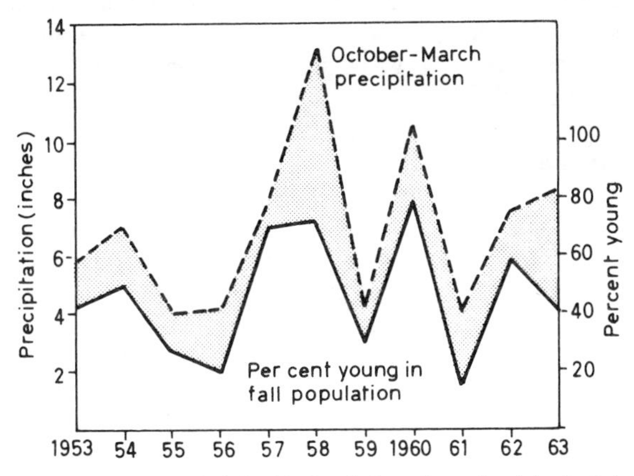

Fig. 4.3 Correlation of winter rainfall with Gambel quail productivity in Arizona. (After Gallizioli [35].)

Disease and parasites can kill many animals. The moose *Alces alces* population in Nova Scotia declined after 1940, as cerebrospinal nematodiasis caused by the parasitic nematode worm *Pneumostrongylus tenuis* increased in frequency. This is carried by the white tailed deer *Odocoileus virginianus* which resists the worm better than the moose does. Hence the ranges of moose and deer now show little overlap in Nova Scotia [36]. Red grouse stocks in some parts of Scotland are affected by louping ill, a tick-borne flavivirus disease, and densities are much reduced compared with virus-free parts of Scotland [37]. The fungus *Entmophthora grylli* has, in one night, killed up to 10% of the nymphal population of the grasshopper *Zonocerus variegatus* in Southern Nigeria [38]. The main known cause of adult mortality was the parasitic fly *Blaesoxipha filipjevi*: 40% or more of the adult grasshopper population died after fly larvae emerged through holes they had eaten in the grasshoppers' dorsal neck membranes.

Predation can account for much of the mortality which a population suffers. Pearson [39] estimated that carnivores ate 80% of a peak population of California voles *Microtus californicus*. Where the sea otter *Enhydra lutris* is abundant, their prey, sea urchins *Strongylocentrotus polyacanthus*, are small and scarce; where sea otters are absent, sea urchins are abundant, large and probably limited by intraspecific competition [40].

Places to live or breed may limit a population. Scaled quail *Callipepla squamata* increased in number after Snyder [41] had added piles of brushwood to an area in Colorado. Saunders and Smith [42] nearly doubled a population of brook trout *Salvelinus fontinalis* by adding physical structures which provided extra territorial stations. Bustard [43] found that the density of gekkos *Gehyra variegata* was less in a forest exploited by man than in a natural forest: trees in the latter had more cracks in their bark which is where gekkos live.

Some animals seem to limit their own numbers below any threshold set by weather, food, disease, predation, parasites or places to live. Before pursuing this topic we shall have to define what we mean by 'limit' more carefully.

4.2 Limitation and regulation

Historically, the idea of limiting factors came from thinking about simple physical or chemical systems. When a chemist adds 1 gram molecule of sodium hydroxide to 2 gram molecules of hydrochloric acid, he gets 1 gram molecule of sodium chloride. The sodium ion is the limiting factor. In this and more complex physical or chemical systems, one material is always in shorter supply than the rest. Only one material can be limiting at any one time. If we add more of the first limiting material to the system then the material in next shortest supply – the second limiting material – becomes limiting, and so on.

Biological systems are more flexible than this. Two or more factors can be limiting at the same time. A man can suffer from both rickets and kwashiorkor. Calcium and protein are both then limiting. Supply of extra protein may ease his kwashiorkor without curing the rickets, or the restriction in performance due to the rickets. Again, lack of food and disease may both reduce an animal's performance without actually killing it.

The classic experimental design used to identify the first limiting factor of a system is to add some of the material thought to be limiting: an increase in performance signifies that the limiting factor has been found. However, in animal populations, such an observation means only that one of the limiting factors, and not necessarily the most important one, has been found. Thus, increasing the animals' food supply may increase their numbers. From this we can say that food supply was one of the factors limiting the population, but not necessarily the only one. By a combination of observation and experiment we compile a list of the important limiting factors. The next step might be to build a mathematical model incorporating our quantitative observations. If the model predicts future population levels successfully, then we understand what limits the population.

From this point of view, we do not regard populations as being 'regulated' about some hypothetical equilibrium density. Instead, limitation of animal populations is defined as that process or those processes which set limits to animal abundance. With this approach, the idea of an 'equilibrium density', developed in Chapter 3, is not regarded as helpful in understanding what controls animal numbers.

The identification of limiting factors is an analytical task. Limiting factors may or may not be related to population density. Once we have enough information about a given problem, probably from several studies, we can make general statements about processes. In turn, this general understanding should help us tackle new problems with greater insight and efficiency.

4.3 Steady populations

If a population remains steady in density over many years, there are no changes in density to explain. We can, however, ask the question: what determines the particular density that we see? The first tool of the population biologist – to correlate population change with variations in some possibly limiting factor – cannot be used this time. Another approach is necessary.

A first step might be to compare densities in different areas. In the west of Scotland, Houston [44] found denser breeding populations of hooded crows *Corvus cornix* in grassy lowland glens than in higher heather moorland. Lowland birds fed their chicks with a wider variety of bigger arthropods than highland ones. A reasonable hypothesis might be that crows are limited by food. This idea needs to be tested, however, because there are many differences between glens and uplands, and the association between crows and food might be due to chance.

Charles [45] studied crows near Ellon: bleak almost treeless farming country in the east of Scotland. Breeding pairs of crows defended the territories in which they nested. They excluded non-breeding crows, which moved around as a flock on treeless land outside the breeding area. When Charles removed territorial crows, flock birds replaced them and bred successfully. When he provided extra trees outside the usual breeding range, flock crows nested in these. But when he increased the number of trees inside the breeding range, this had no effect.

What can we learn from these experiments? First, we have to be careful when we use the word population. There was a breeding population and a non-breeding population. Second, the experiments teach us to distinguish number from density. Expanding the area with trees expanded the breeding range, and it would be quite reasonable to conclude that trees limited the number in the breeding population. However, as extra trees inside the breeding range had no effect, trees did not limit breeding density in this population. *What would you have concluded about the effect of trees on breeding density if Charles had chosen a much larger study area including several breeding units and several non-breeding flocks?*

In another experiment, Yom-Tov [46] put out extra food in the form of hens' eggs and chicks. The crows ate the food but kept the same territories, and so population density did not increase and was not limited by the immediate availability of food. The crow, however, is a long-lived bird and it is natural to speculate what might have happened had Yom-Tov kept feeding the birds for many years. The birds might have held on to their traditional boundaries for some time, after which they or their successors might have taken smaller territories. If so, the birds' territorial behaviour would have been implicated as a factor limiting population density in the short term and food in the long term.

Charles [45] tested this idea directly by implanting testosterone pellets into flock crows. They became more aggressive and persistently intruded into the breeding range, finally setting up extra territories. Breeding density was limited by territorial behaviour.

Many vertebrate predators and carrion eaters resemble the crow. Populations are steady in density for years but comparisons between areas show higher densities where there is more food. The best explanation is that territorial behaviour is the factor limiting abundance in the short term, but that the size of territories can be modified by the availability of food over many years. Such animals are *K*-selected and do not track (Section 2.3) short-term fluctuations in the environment – such as the eggs and chicks provided by Yom-Tov – but do adapt to long-term changes.

In such cases it is useful to introduce another logical tool, the distinction between 'proximate' and 'ultimate' limiting factors. Proximate factors are those which can be seen to have an immediate effect, such as territorial behaviour in the crow. However, food supply might have a long-term effect on territory size and so might be an ultimate limiting factor. Again, starvation might weaken an animal and so predispose it to predation. The proximate cause of death would be predation, but the ultimate cause starvation. Proximate and ultimate are only relative terms; to the student of evolution, the ultimate cause of much of what we see about us is natural selection.

4.4 Fluctuating populations

Many populations of small rodents such as voles and lemmings undergo periodic peaks and crashes in population density about once every three or four years. It is tempting to assume that such cycles are due to one overriding process, and many workers have tried to discover what this might be.

Pearson [47] and Krebs [48] studied cyclic populations of the California vole within a few hundred yards of each other and reported very similar densities (Fig. 4.4). Pearson was impressed by the impact of predators on the population during and following the decline of the vole population (Fig. 4.5). As vole numbers fell, so did the number of predators, but not so fast. Hence the ratio of predators to voles increased and the proportion of the vole population eaten was at its peak when vole densities were at their lowest. Similar patterns were observed in two successive declines. Pearson concluded that predation was likely to be responsible for the timing (period) and amplitude of the cycle in California and elsewhere. He was not so sure that carnivores were responsible for the initial decline from high density, but thought that this might be so.

Krebs, however, noted little predation on some of his study plots, yet these areas also showed fluctuations in vole numbers. Furthermore, he found delayed breeding to be characteristic of declining populations and lengthened breeding seasons in increasing populations. This was unlikely to be caused by predation. Food, too, seemed unimportant: providing a declining population with extra food failed to halt a decline. Disease is sometimes associated with declines: epizootics of tularemia can occur in dense populations of voles. But this is not always the case

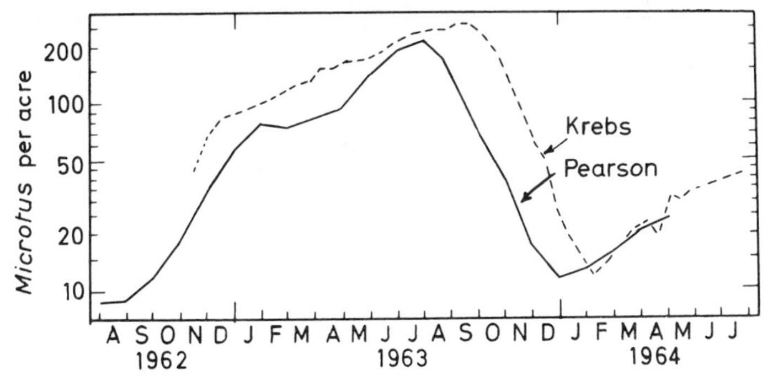

Fig. 4.4 Comparison of the population densities of California voles reported by Pearson (1966) and Krebs (1966) on adjacent study areas. (After Pearson [47].)

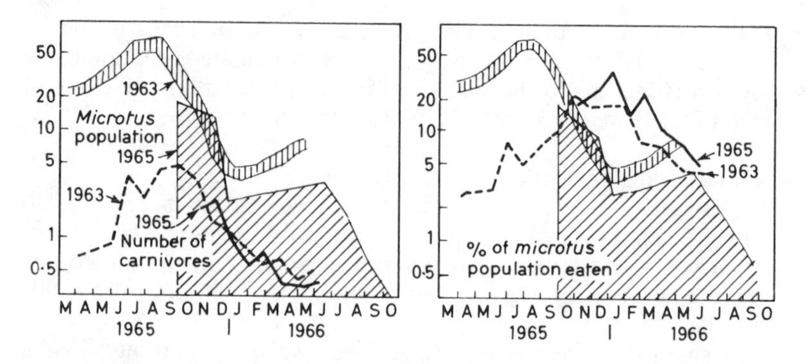

Fig. 4.5 Comparison of the number of carnivores and the number of voles ($\times 10^{-2}$) present on 35 acres at comparable stages in the decline of the 1963 and 1965 populations of California vole; and, percentage of the vole population, present at the beginning of each month, that carnivores ate in that month. (After Pearson [47].)

and Krebs concluded that disease was unimportant in his study. Neither he nor Pearson had studied behaviour, and this seemed to be the remaining possible cause of cyclic declines.

If there is one process that underlies vole cycles, then some sign of it ought to be present in all cyclic fluctuations. It seems likely, however, that even if there is such a process, other factors are likely to cause declines from time to time, declines divorced from any cyclic process. In analysing the problem, the first step is to characterize the symptoms of a typical cyclic fluctuation. For small rodents, these include the variations in the length of the breeding season noted above, and the appearance of unusually large animals at the peak of the cycle. The next step is to list those factors present in each cycle. If heavy predation is present during some but not all cyclic declines, then predation is not a necessary factor in a cyclic decline. Starvation, predation and disease may be sufficient to cause vole numbers to decline on some occasions, but Krebs' work suggests that they are not necessary. It does not, however, show that there is a necessary factor: that is, a proximate factor common to all declines.

An alternative view is that once voles build up to high densities, this predisposes them to heavy mortality from all kinds of factors. On one occasion this might be disease, on another predation, on a third starvation and on a fourth physiological stress resulting from over-crowding [49]. Hence there need be no necessary factor. The only consistent process might be a rapid build up from low numbers after one decline and preceding the next.

The 'holistic' view that many factors affect animal densities is unassailable but woolly. It may lead a worker to collect data on as many factors affecting a population as he can. Such studies tend to produce statistical correlations, but little improvement in understanding. In contrast, attempts to detect a particular process in action often lead to experimental tests of a particular hypothesis. The experiment may disprove the hypothesis: if so, we have learned something useful.

Can we find a process that is always present in cyclic vole declines? Krebs' view is that this is self-regulation: changes in the animals' spacing behaviour cause changes in density independently of changes in the environment.

4.4.1 Dispersal and self-regulation

The breeding population of crows studied by Charles [45] was steady in number from one year to the next and limited by territorial behaviour. Most vertebrate zoologists would be happy to call this self-regulation without further thought, but an entomologist might ask to be shown that territorial behaviour is density-dependent. When flock birds intruded into the breeding area the intensity of agonistic and territorial behaviour increased markedly. Hence the total density of the birds had increased along with the intensity of territorial behaviour: the latter was indeed density-dependent and the use of the word regulation in its narrow sense (Section 3.3.2) is justified here.

It seems reasonable that a territorial bird with steady breeding densities should be self-regulated. For a long time, however, people found difficulty with the idea that territory size could vary from year to year and still be limiting population density each year. A view of the territory as a rubber disc of more-or-less fixed area hindered thinking in this field. Lack [50] even used population steadiness as a criterion for distinguishing between birds whose population was limited by territorial behaviour and those where other factors were limiting.

The red grouse shows marked fluctuations in density, peaking about every six years [51]. Each year, young and old cocks compete for territories in autumn and hold them until the breeding season. By then, cocks which fail to get a territory and hens which fail to get a cock are almost all dead. All the available habitat is occupied by territorial birds and so territory size varies inversely with the density of the breeding population. As in the crow, non-territorial birds will take territories when their aggressiveness is artificially boosted with an implant of testosterone. Cocks which already have territories and are implanted with testosterone increase their area at the expense of neighbours (Fig. 4.6). Territorial behaviour limits the population each year and changes in mean territory size cause changes in population density from one year to the next.

Every year, red grouse which fail to join the breeding population roam more widely than territory owners, before they die from a variety of proximate causes. The only necessary cause of death is not having a territory. Cyclic declines are characterized by mass emigration from the population and long-distance dispersal [53].

Self-regulation in steady populations too is accompanied by dispersal. Grey squirrels *Sciurus carolinensis* in Mount Pleasant cemetery in Toronto produced more than enough young to compensate for the low

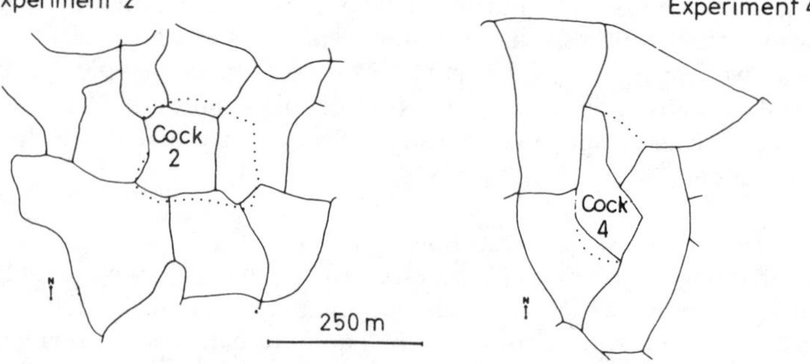

Fig. 4.6 Increase in the territory size of cock red grouse after they were implanted with testosterone. Dotted line indicates new territorial boundaries of implanted cocks. (After Watson and Parr [52].)

rate of adult mortality [54]. Young which failed to establish themselves in the population dispersed away from the cemetery.

Krebs, Keller and Tamarin [55] fenced in a fluctuating population of field voles so that they could not disperse. Numbers built up to unprecedented levels until they had eaten much of the grass in the fenced area and the population declined. On a control area, peak densities were much lower and the population declined well before that in the exclosure, without obvious over-grazing. On the control area, much dispersal had occurred, as is typical both of fluctuating populations of *Microtus* spp. and of steady populations like the grey squirrel.

Self-regulation alone cannot explain cycles in population density because it also occurs in steady populations. Furthermore, some dispersal occurs from all populations. The pattern, however, may differ between populations and a cyclic pattern of dispersal may characterize cyclic population fluctuations (Section 4.4.3).

4.4.2 Dispersal and predation

Tamarin [56] studied dispersal in island and mainland meadow voles *Microtus* spp. The mainland voles showed fluctuations and dispersal patterns typical of the usual microtine 3–4 year cycle, but the island voles remained more or less steady in numbers and did not disperse in the usual pattern. There were no mammalian predators on the island and so both Pearson's [47] idea that predation is necessary to cycles, and Krebs' idea that dispersal resulting from self-regulation is necessary, were confirmed (i.e. the evidence was consistent with the hypothesis).

Tamarin concluded that dispersal was necessary to cycles. Self-regulation would result in the less dominant animals leaving the population and becoming more vulnerable to predation as a result. Dispersing water voles *Arvicola terrestris* are more likely to be eaten by herons and owls than residents [57] and non-territorial red grouse are also more vulnerable to predation than territory owners [58]. The idea of increased mortality amongst a socially subordinate part of the population seems to be a reality, but probably occurs in both steady and fluctuating populations.

In some species, the subordinate animals are doomed to die as a result of having low status. If a predator does not eat them, and if they are not killed by disease or parasites multiplying in their stressed bodies, then they will die anyway of stress alone. The social system by itself may be enough to kill them – this can occur in the red grouse. On the other hand the die-offs seen in many deer populations may be the result of subordinate animals surviving in the absence of natural predators. If predators capable of removing subordinate animals are abundant enough, they may regulate deer numbers in conjunction with the social system, so that excessive densities and subsequent die-offs do not occur. In Europe the roe deer often lives in isolated woods surrounded by farmland: the old bucks and does hold territories in the woods and expel young animals when they reach maturity. Predation by human beings removes the young animals from the farmland. When fenced, roe deer increase in numbers until signs of heavy grazing are evident on the vegetation and the animals are smaller and breed less well than on the unfenced areas [3]. They would probably be much more likely to die than control animals in the event of heavy snow, but this has not been shown.

4.4.3 Dispersal and selection

The important role of dispersal in self-regulation and the possibility that this might result in genetic selection for different demographic characteristics is one possible way of explaining apparently causeless fluctuations in animal numbers. Equally possible in our present state of knowledge is that dispersal and other aspects of population change might involve selection for particular genotypes, but that the genotypes themselves have no effect on demography.

However, laboratory observations show clearly that populations started with an identical number of founders in similar environments can achieve quite different population densities [59]. This is partly attributable to chance, but partly to genetic differences between populations. It seems to follow that selection for genotypes which result in different population densities should be possible.

Adjacent populations in the wild can show marked genetic differences, and dispersing blue grouse *Dendragapus obscurus* and voles *Microtus* spp. seem to differ genetically from residents [60, 61]. One problem with work on wild animals is that the demonstration of such genetic differences has usually depended on the use of genetic markers, such as blood proteins, which are of unknown relevance to population dynamics. Greenwood, Harvey and Perrins [62], however, showed that the distance which dispersing great tits *Parus major* move is a heritable trait.

One idea for explaining cyclic fluctuations is to suggest the existence of two genotypes, an aggressive, dominant and resident type, and a subordinate, unaggressive and dispersing type. The dispersing type is likely to found new populations and to be at no disadvantage at low densities when contacts between neighbours are few. As numbers build up, one may speculate that the environment becomes saturated with animals, and the workings of the social system ensure that the subordinate dispersers move out and leave the dominant residents. As the proportion of dominant residents in the population increases, so disputes between them become more frequent and this might interfere with breeding and result in a decline in population density. The precise details of the mechanism might vary, depending on whether the dispersers left the population in greatest numbers before the peak in density (as in microtines) or after it (as in red grouse) (Fig. 4.7). In both cases, the prediction first made by Chitty [63] would be that a high proportion of inherently aggressive animals should remain at the end of the decline. This seems to occur in red grouse [64].

4.5 Social organizations and population density

An essential role of any social system from the viewpoint of population dynamics is that it spaces animals out. The kinds of behaviour which cause the observed spacing pattern can be called spacing behaviour. The precise details of social structure, whether the system is territorial, tribal, hierarchical, are less important in determining numbers than their effects, such as dispersal.

There have been three views about the importance of spacing behaviour in causing changes in population density. It may be unimportant. Territorial behaviour, from this standpoint, does not limit animal numbers. It simply spaces them out after they have been limited by some other resource such as food. The socially dominant animals may well be the ones which survive competition for food, but social structure has no role to play in determining how many animals survive. Animal numbers can be modelled and predicted without recourse to studies of spacing behaviour.

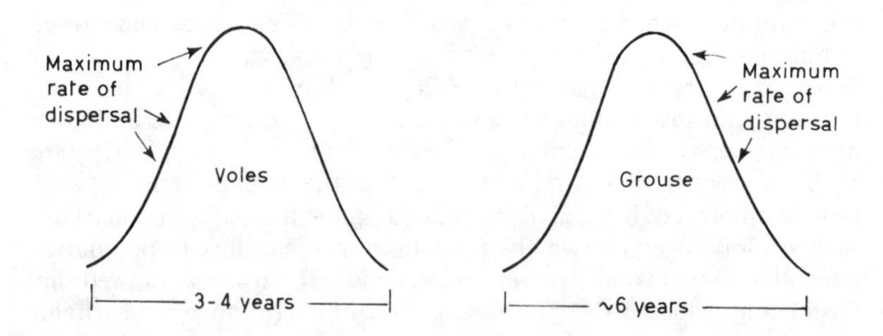

Fig. 4.7 Dispersal in voles *Microtus* spp. and red grouse, in relation to the phase of the cycle.

Another opinion takes note of the fact that many animal populations vary in parallel with food supplies, but do not seem to be limited by food in the short term. An example of this is the crow (Section 4.3). The observation that there are more animals where there is more food is commonplace; but examples of starvation are much rarer. One explanation is that animals regulate their own numbers below any limit set by food shortage, that the social system acts as a buffer against starvation. At the same time, animals detect how much food there is and vary their numbers in parallel with supplies [65].

This is not so difficult as it sounds. A herbivore, for example, has a range of foods of varying quality available to it. As numbers increase, so the quality of the diet decreases; body size and breeding performance too might decline. Such a decline in food quality might trigger changes in behaviour which repress breeding and initiate dispersal well before obvious damage to the food supplies can be seen.

A third view is becoming increasingly popular as environmental changes seem less and less adequate to explain all observed changes in animal numbers. This view does not deny the importance of environmental factors. At the same time, the tendency of animals to space themselves and thus regulate their own density may operate on its own, with scant effect from the immediate physical environment.

There is no consensus. Some workers staunchly ignore behaviour and find that they can explain their observations without it. Others regard social organization as an important mechanism linking population densities to the environment, but changing only in response to changes in the environment. Showing that behaviour can affect numbers independently of the environment has been done by implanting testosterone into the crow and the red grouse. Showing that it actually does so in natural populations is more difficult because it means ruling out all environmental variables, and critical colleagues will always think of something not measured in a particular study. The conditions that need to be met are outlined in Table 4.1.

Probably, each of the three views is correct for some populations and incorrect for others. As with the conflict between workers who studied

Table 4.1 Conditions which will show that behaviour, via socially-induced mortality (or socially-induced depression of recruitment), limits a breeding population. (Simplified from Watson and Moss [25].)

(a) A substantial part of the population does not breed, either because animals die; or because they attempt to breed but they and/or their young all die; or because they are inhibited from breeding even though they survive, and may breed in later years.
(b) Such non-breeders are physiologically capable of breeding if the more dominant or territorial (i.e. breeding) animals are removed.
(c) The breeding animals are not completely using up some resource, such as food, space or nest sites. If they are, the resource itself is limiting.
(d) The mortality or depressed recruitment due to the limiting factor(s) changes in an opposite sense to other causes of mortality or depressed recruitment.

central populations and believed in density-dependent regulation, and those who studied edge populations and emphasized the importance of environmental factors (Section 3.3.4), a synthesis will emerge and polemics be seen to be due to partial understanding.

Such a synthesis will come sooner if we learn to distinguish between phenomena and scholarship. A careful appraisal of the available facts, rather than of what eminent workers have written about these facts, will lead to better understanding of what determines animal numbers.

5 The natural history of numbers

Having studied an animal population for some time, we end up with observations couched as a set of numbers. We might have estimates of population size, birth rates, mortality rates and so on. We notice relationships between these numbers, notably that births and immigrations minus deaths and emigrations equal change in population size. We might be tempted to model our population in these terms. The bathtub model (Fig. 5.1) immediately suggests two things: first, gains and losses should be expressed as rates analogous to the rates of flow of water; second, once we know the correct mathematical relationship between gains and losses, we should be able to predict population size in the future.

The difference between gains and losses is called the net rate of change of the population:

$$\mathrm{d}P(t)/\mathrm{d}t = i(P,\ t) - m(P,\ t) \qquad (5.1)$$

In this chapter the equations are all explained in words and most in graphs, both to ease understanding and to illustrate the crudeness of the ecological assumptions that they represent.

In equation (5.1), the net rate of change of the population $\mathrm{d}P(t)/\mathrm{d}t$ equals the rate of increase $i(P, t)$ minus the rate of decrease $m(P, t)$. When we write $i(P, t)$ we mean that there is a formula, called i (based on P the amount of water in the bath or the population size, and t, the time) that describes the rate that water is flowing into the bath or the rate of gain to the population, at time t. More briefly, i is a function of P and t. Similarly, the rate of loss is described by m (which is a different function of P and t). Equation (5.1) is just a general statement: we do not know

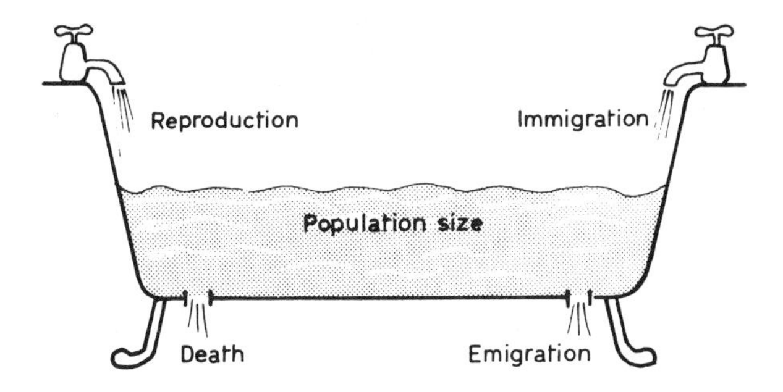

Fig. 5.1 A bathtub with two taps and two plugholes.

the hoped-for formulae represented by i and m, and much work in theoretical population dynamics is an attempt to find them.

Any prediction of population size at time t will need to include one measurement of the amount of water in the bath at some time, say P_0 at time 0. Such a prediction can be made by solving equation (5.1), which is a differential equation. The solution will be a new function, called f, containing P_0 and t.

$$P(t) = f(P_0, t) \tag{5.2}$$

5.1 Exponential growth

In equation (5.1) we made i and m functions of P. Here the bathtub analogy gets strained because there is no reason why the amount of water flowing in should be related to the amount already there. In an animal population, however, the more animals there are, the more offspring they can produce. The net number of young produced per animal per unit time can be called the 'net reproductive rate'. This implies that we tacitly ignore death and movement in and out of the population, or else use a definition of 'reproduction' which includes birth, death and movement. The rate of growth of our population of P individuals is rP, where r is the maximum possible net reproductive rate.

$$\frac{dP}{dt} = rP \tag{5.3}$$

The term on the left hand side is read as 'dP by dt' and means 'the net rate of growth of the population' (Fig. 5.2). It looks like a fraction but the top and bottom bits cannot be separated. Apart from this, the equation can be worked according to the usual rules of algebra. Thus we can write

$$\frac{1}{P}\frac{dP}{dt} = r \tag{5.4}$$

which states that the left hand side, which is called the 'specific growth rate', is constant and equal in this model to r (Fig. 5.3). The specific growth rate of a population is the rate of population growth per individual animal. A population of P animals growing at a rate dP/dt has a specific growth rate of $1/P\ dP/dt$.

To predict P at time t, we solve equation (5.3) to get

$$P(t) = P_0 \exp(rt) \tag{5.5}$$

where $P(t)$ is P at time t and exp means 'the exponential of'.* This is the equation for exponential growth (Figs 5.4 and 5.5). Exponential growth can occur in populations at low density in favourable environments, but sooner or later the net reproductive rate and population growth begin to decrease.

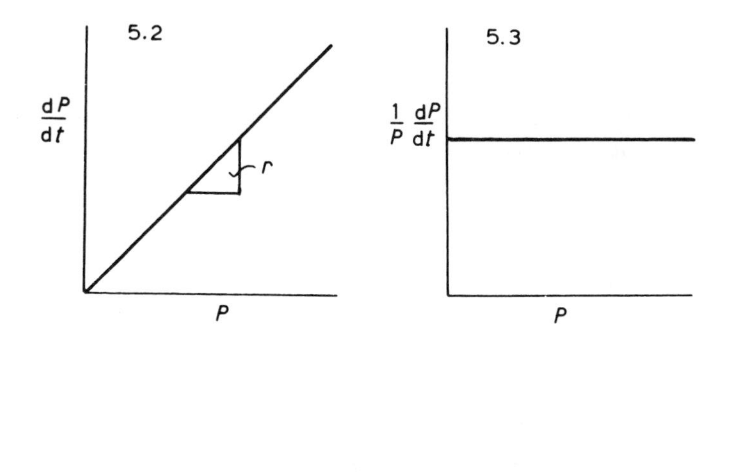

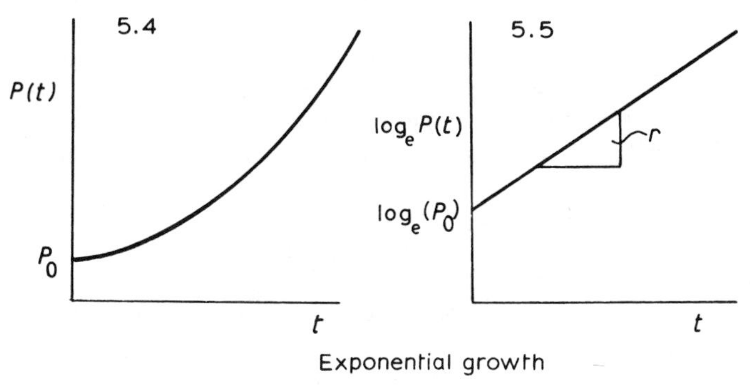

Fig. 5.2 Linear relationship with a slope of r between the growth rate of the population and population size. The larger the population, the faster it grows.
Fig. 5.3 The specific growth rate of the population remains constant and equal to r at all population sizes.
Fig. 5.4 The population grows faster as time passes.
Fig. 5.5 When the population size is expressed as a logarithm, the curve of Fig. 5.4 is transformed into a straight line with a slope of r.

* $\exp(x) = e^x$
where e is a constant number (2.718) with special properties. The two sides of this expression mean exactly the same thing. To obtain a value of $\exp(x)$ for a particular x substitute that value into the expression. Thus $\exp(0.5) = e^{0.5} = 2.718^{0.5} = 1.65$. The solution of the differential equation

$$dP/dt = rP$$

is
$$P(t) = P_0 \exp(rt).$$

The graph of $P(t)$ plotted against t is shown in Fig. 5.4. Taking logarithms to the base e

$$\log_e(P(t)) = \log_e(P_0) + rt \log_e(e)$$
$$= \log_e(P_0) + rt$$

This is the equation of a straight line with slope r (Fig. 5.5).

5.2 Logistic growth

A possible assumption is that there are effects inhibiting growth and proportional to population density, so that the population's specific growth rate decreases as density approaches the maximum number of animals that the environment can support, which is K and often determined by available resources. In symbols

$$\frac{1}{P}\frac{\mathrm{d}P}{\mathrm{d}t}=r\left(1-\frac{P}{K}\right) \tag{5.6}$$

as illustrated in Fig. 5.6.

Equation (5.6) can be rewritten

$$\frac{\mathrm{d}P}{\mathrm{d}t}=rP\left(\frac{K-P}{K}\right) \tag{5.7}$$

and the resulting relation between growth rate and population size is shown in Fig. 5.7. The term in the brackets is what makes this equation different from the exponential growth equation (5.3) and modifies the net reproductive rate in a density-dependent way. When P is small, the term in brackets is close to 1 and growth nearly exponential; but when P approaches K, the term is close to zero and so is the growth rate of the population.

The term rP outside the brackets is the same as the right hand side of the exponential growth equation (5.3) and is sometimes called the 'exponential growth term'. The term inside the brackets diminishes the exponential term in proportion to density and hence can be called the 'density-dependent term'. In doing so it limits the number of animals to K, the 'carrying capacity' of the environment. At K the environment is supplying just enough food and other requirements to maintain the population.

The type of growth described by equations (5.6) and (5.7) is called 'logistic growth'. Solving the differential equation (5.7) gives another equation that describes the logistic growth of a population as a function of time. The essential features of this equation are illustrated in Figs 5.8 and 5.9.

5.2.1 Logistic growth and real animals

The logistic growth equation is derived from purely theoretical considerations. How well does it describe the growth of real animal populations? To answer this, the first step is to find if the equation is adequate when all its underlying assumptions are met, and this can be done only in the laboratory.

We need a constant environment with a fixed carrying capacity or K. This means that we have to supply living requirements at a rate sufficient to maintain a fixed population density. Early workers interested in the general problem of population growth selected organisms such as yeast cells; these could become dormant when no longer able to reproduce and

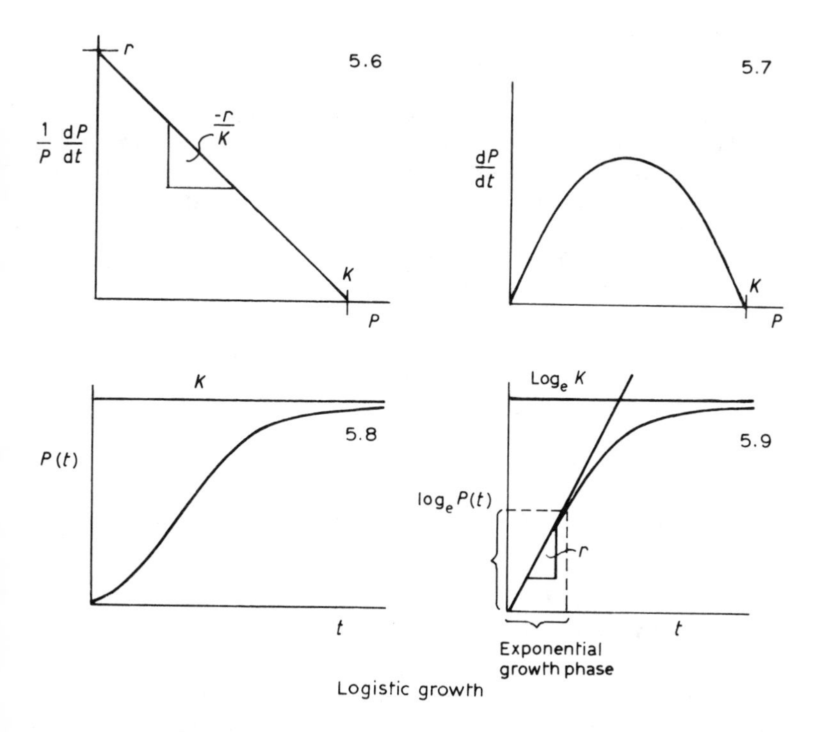

Fig. 5.6 Decreasing linear relationship between the population's specific growth rate and its size. Contrast with Fig. 5.3. When P is close to zero, the specific growth rate is close to r and growth approximately exponential. When the population reaches K it ceases to grow. The line is thus defined at two points and its slope is $-r/K$.
Fig. 5.7 The growth rate of the population at first increases and then decreases as the population grows in size. The fastest growth rate is when the population is half its maximum size.
Fig. 5.8 Population size increases quickly at first and then more slowly as it approaches K.
Fig. 5.9 When population size is expressed as a logarithm, the curve of Fig. 5.8 looks like this. At small P, growth is approximately exponential, but as the population approaches K, growth departs further and further from the exponential model.

then had essentially zero maintenance requirements. The logistic growth equation fitted growing populations of yeast cells in a batch of nutrient medium quite well. Most animals, however, do have maintenance requirements.

According to the logistic growth model, the specific growth rate of the population decreases linearly with population density for a given K, but the model says nothing about the actual mechanism. In a laboratory culture of organisms of a given species and stock, it is reasonable to assume that the amount of food available to each individual might limit its reproductive rate and so the growth rate of the population. If so, the ratio of food to organism should determine the population's growth rate.

Smith [66] fixed different levels for the specific growth rates of several populations of the water flea *Daphnia magna* in different tanks, equivalent to different points on a population growth curve. Each day he supplied food and increased the volume in each tank by adding fixed percentages of a suspension of bacteria and algae. Eventually the populations in the tanks stabilized at different densities, when the growth rate of each population equalled the rate at which the suspension was added. When small percentages of suspension were added, the populations reached equilibrium at high densities with low growth rates, when the rate at which animals were diluted equalled the population's growth rate. When large percentages were added, equilibrium occurred at low densities with high growth rates.

According to the logistic model, the specific growth rate of a population should decrease linearly with population density. Smith found that the growth rate declined more rapidly with population density than the logistic model would predict (contrast Figs 5.6 and 5.10). To take account of this, he changed the density-dependent term in the logistic equation (5.7)

$$\frac{\mathrm{d}M}{\mathrm{d}t} = rM\left(\frac{K-M}{K+r/cM}\right) \tag{5.8}$$

He preferred to use M, the biomass of the animals, rather than P, the number. The constant c is equal to the rate of replacement of biomass per unit mass present in the population when it has reached carrying capacity. This modification can be interpreted by saying that if an animal is breeding rapidly its maintenance requirements increase, irrespective of its requirements for producing offspring.

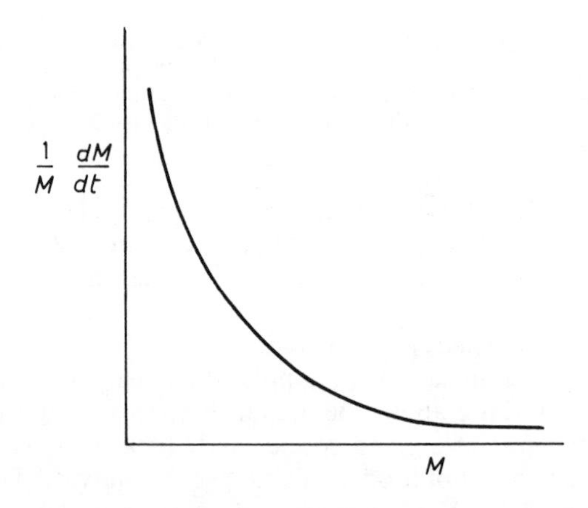

Fig. 5.10 Form of the observed relationship between the specific growth rates of populations of *Daphnia magna* and their densities. It is markedly concave. Contrast with Fig. 5.6 (After Smith [66].)

Smith set up his experiments to measure the specific growth rate of several populations and was unable to measure K directly in all of them. *Does this affect your interpretation of his results?*

5.2.2 Logistic growth and time lags

Logistic growth may be a poor mimic of reality, but is simple and therefore favoured by theoretical ecologists. One way of accounting for fluctuations in animal populations is to point to the impact that they have on their own growth rate. Thus deer are thought to build up in numbers until they surpass the carrying capacity of their range and then starve. In terms of the logistic equation (5.7), P has exceeded K by enough to cause the growth rate of the population to become negative. Deer numbers then decline to well below the carrying capacity. A simple model is to pretend that the overpopulation at time $t-T$ reduces the growth rate of the population at time t. For example, we might suggest that a hind poorly fed at time $t-T$ is less likely to produce a calf at time t. The size of this effect is likely to be proportional to the extent by which P exceeded K at time $t-T$. It is as if the density of the population some time ago were affecting the growth rate now.

$$\frac{\mathrm{d}P(t)}{\mathrm{d}t} = rP(t)\left(\frac{K-P(t-T)}{K}\right) \tag{5.9}$$

Here we have replaced the P in the exponential term of equation (5.7) by $P(t)$, which means the density of the population at time t. The P in the density-dependent term has been replaced by $P(t-T)$, the density of the population T time units before time t. In other words, the regulation of growth is lagging behind the density that caused it, by a time lag of T. This is a model of a 'delayed density-dependent' (Section 3.3.3) relation between growth rate and population size.

This and related time-lag models have a fascinating array of properties [67]. When T is zero we revert to equation (5.7), or simple logistic growth. Other results depend on the exact value assigned to rT: model populations can become extinct, oscillate with decreasing amplitude until a stable population density is achieved, or oscillate regularly and indefinitely.

The dynamics of time-lagged logistic models depend entirely on rT. K may determine mean population densities, but has no effect on the pattern of the fluctuations.

Time-lag models can generate the main features of most population fluctuations. An enthusiast might suggest that it is only necessary to measure r and T, and that this would be sufficient understanding to allow us to explain and predict most population changes. *How would you respond to this suggestion?*

5.2.3 r*-selection and* K*-selection*

We are now in a position to understand these terms. They refer in broad terms to the logistic growth equation. Animals which are r-selected have a high r: birth rates can exceed death rates by an amount sufficient to allow rapid increase in the population. K-selected animals have a low r and specialize in maintaining population densities which do not exceed K. They often have stable densities because life spans are generally long; in general, their numbers are well-regulated in the narrow sense of Section 3.3. Populations of K-selected animals are unlikely to become extinct as long as they are in the environment to which they are adapted. When the environment changes catastrophically, K-selected animals are in danger of extinction.

Populations of r-selected animals might well overshoot K because they have a high r. Whether they actually do so might also depend on the time lag T. A long T will mean that regulation will not operate until after P has exceeded K. Hence r-selected animals might be more likely to deplete and damage their source of food, more likely to oscillate and also more likely to become extinct. They can afford to take this chance because they often affect the type of life-history that we earlier termed survival by dispersal (Section 3.3.1).

Many more analogies between the logistic equation and animal life-styles can be drawn [68]; but they are only analogies. The human brain tends to think in dichotomies and this is one of them. The terms r-selected and K-selected are useful shorthand for conveying two ends of a poorly-defined spectrum, and no more than that.

For example, the carrying capacity of a given piece of ground is a theoretical notion which is usually difficult to substantiate. In animals which are regulated by social behaviour, the carrying capacity of the environment may play little or no role in determining numbers.

5.3 Organisms in chemostats

Many uncertainties in the real world can be avoided in the chemostat (Fig. 5.11). This microcosm contains a constant volume of a suspension of micro-organisms, usually of one species. Micro-organisms are not animals but one hopes that they follow some of the same rules as animal populations. A solution of nutrients is added to the chemostat and an equal volume of suspension taken away at the same rate. The nutrient solution is of known composition, and supplies all the nutrients needed for life in abundance, except for one, the limiting nutrient. The withdrawn portion of the suspension contains organisms, part-used nutrient solution, and wastes. Gains to the population depend on the reproduction of the organism. Losses occur as organisms are swept out of the vessel. Food supplies are regulated by adjusting the concentration of the limiting nutrient.

Ideas about population growth can be tested in the chemostat. If they work here, they may apply in the field. In this treatment, taken from Herbert, Elsworth and Telling [69], we discuss the relationship between the observed growth rate of a population, and the concentration of the limiting nutrient in its environment. In theory, the growth rate should decrease as the quantity of limiting nutrient per organism declines.

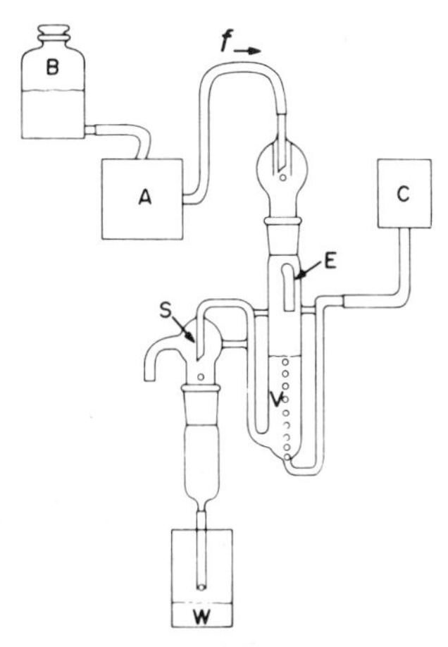

Fig. 5.11 A chemostat. A peristaltic pump (A) pumps fresh medium from the reserve flask (B) into the chemostat at a set flow rate, *f*. The cells grow in the liquid, V. For each drop of fresh medium added, a drop is siphoned off (S) and the cells and used medium go into collecting flask (W). An aquarium pump (C) is used to aerate and mix the culture. A large covered port on the side of the chemostat (E) allows the air to escape, preventing a pressure build up. (After Dykhuizen and Davies [70].)

We start with the usual statement that gains minus losses define population change; in this case we use rates of change

$$dP/dt = \mu P - DP \qquad (5.10)$$

where P is the density of organisms, μ the specific growth rate of the population and D the dilution rate. D is the number of changes of volume per unit time, a convenient description of the flow rate of the system (f/volume of liquid V in Fig. 5.11).

Observations show that the specific growth rate of micro-organisms depends on the concentration of the limiting nutrient

$$\mu = \mu_m \left(\frac{S}{K_s + S} \right) \qquad (5.11)$$

where μ_m is the maximum specific growth rate, S the concentration of the limiting nutrient in the chemostat, and K_s the saturation constant. The saturation constant is the concentration of limiting nutrient that produces a growth rate of $\frac{1}{2}\mu_m$. As with Smith's work on *Daphnia* (Section 5.2.1), the growth rate is usually measured when the population density is constant; then the growth rate is equal to the dilution rate D set by the experimenter.

How does this compare with the logistic equation? In form, it is not dissimilar to equation (5.6), with μ_m analogous to the exponential growth rate r, and with the term in brackets modifying the growth rate. In both cases, growth rate decreases with increased population density, in equation (5.6) simply by definition; and in equation (5.11) because the concentration of the limiting nutrient is likely to be less with denser populations (Fig. 5.12). The virtue of equation (5.11) is that all the parameters can be measured and so the model tested against observed numbers.

Substituting equation (5.10) in equation (5.11), we get

$$\frac{dP}{dt} = \mu_m P \left(\frac{S}{K_s + S} \right) - DP \qquad (5.12)$$

Here we see a difference from the logistic equation. In equation (5.12), losses are explicitly modelled as the rate of loss of organism-containing medium, whereas the logistic equation ignores losses and we have to pretend that the reproductive rate includes both gains and losses. The loss term DP is density-independent and cannot have a regulating (in the sense of Section 3.3.2) effect on the population.

The logistic equation assumes that population density has an effect on growth rate, without specifying a mechanism. Now that we are dealing with a real microcosm, we find that the creature is affecting its own environment by removing nutrients. Hence S varies and we have to model this too.

We start by assuming a constant yield Y of organism per unit volume per unit of nutrient consumed:

1 g nutrient $= Y$ g organism
$1/Y$ g nutrient $= 1$ g organism

If the rate of gain to the population is μ, then the organism is removing nutrients at the rate $\mu P/Y$ g per unit time; μ is the specific growth rate (equation (5.11)).

We are adding nutrient to the chemostat at a rate DS_R, where S_R is the concentration of nutrient in the fresh medium. Nutrient is removed at a rate DS, where S is the concentration of nutrient in the part-used medium. The net rate of change of nutrient concentration in the chemostat is the rate at which it is put in, minus the rate at which the

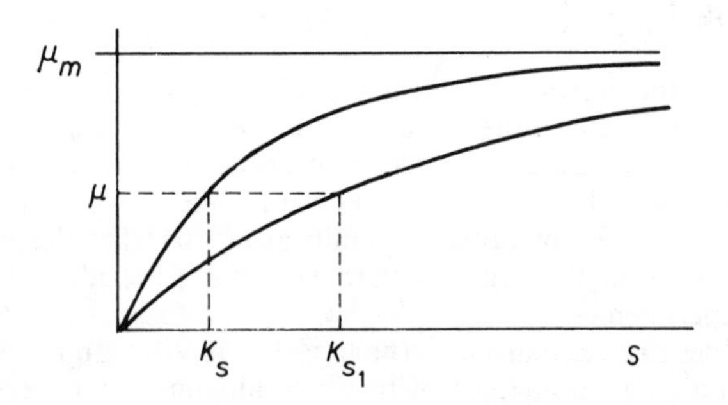

Fig. 5.12 The specific growth rate of a population of micro-organisms and the concentration of the limiting nutrient. The shape of the graph depends on μ_m and the saturation constant K_s. Large K_s means greater S is required to produce a given μ. Two curves with two different values of K_s are shown.

organisms use it up, minus the rate at which it is taken out as part-used medium.

$$\frac{dS}{dt} = DS_R - \frac{1}{Y}\mu_m P\left(\frac{S}{K_s + S}\right) - DS \qquad (5.13)$$

To predict the size of the population, the next step is to solve the pair of equations (5.12) and (5.13). The average biologist will go to a mathematician for help at this stage. The solutions produced by the mathematician will be of biological interest only if the assumptions on which the equations rest are biologically sound. One assumption that might be questioned is that the yield of organisms per unit of nutrient is constant independently of μ. Smith's work on *Daphnia* suggested that it might not be. If the model makes predictions that are accurate enough for our purposes, then the assumptions are acceptable. If the predictions are not accurate enough, then we can refine the model and start again.

5.3.1 The chemostat and an ecological problem

Animal numbers are limited not only by their physical environment and population density, but also by competition for resources with other animals. It is often observed that some animals are specialists and others generalists; the specialist might concentrate on one kind of food whilst the generalist eats a variety, often including the specialist's favoured item. How can both species co-exist and why does one not out-compete the other? A usual answer is that the specialist is more efficient at using its particular food and beats the generalist in competition for it; but the generalist can eat other foods when the shared food is in short supply. This is an excellent *post hoc* explanation of observed patterns, but is very difficult to test in the field.

Dykhuizen and Davies [70] explored the circumstances in which generalist and specialist strains of the gut bacterium *Escherichia coli* could co-exist in the chemostat. The limiting nutrients were the two sugars maltose and lactose. The generalists could utilize both, but the specialists were unable to use lactose. Inefficient generalists and efficient specialists co-existed at roughly similar densities when 5–25% of the sugar solution comprised lactose. However, the specialists out-competed the generalists when there was only 1% lactose, and at 50% lactose the generalists predominated (Fig. 5.13).

The 5% lactose curve was particularly informative. Here the generalists initially declined in density but after about 20 generations suddenly improved their performance, recovering from 1% of the population to stabilize at about 10%. The generalists had adapted. A genetic mutation had occurred which allowed the generalists to use low concentrations of lactose more efficiently than previously. They had evolved. This experiment not only confirmed the idea that generalists and specialists could co-exist despite direct competition for a shared resource. It also showed that genetic changes with considerable effects on demography can occur in a few generations.

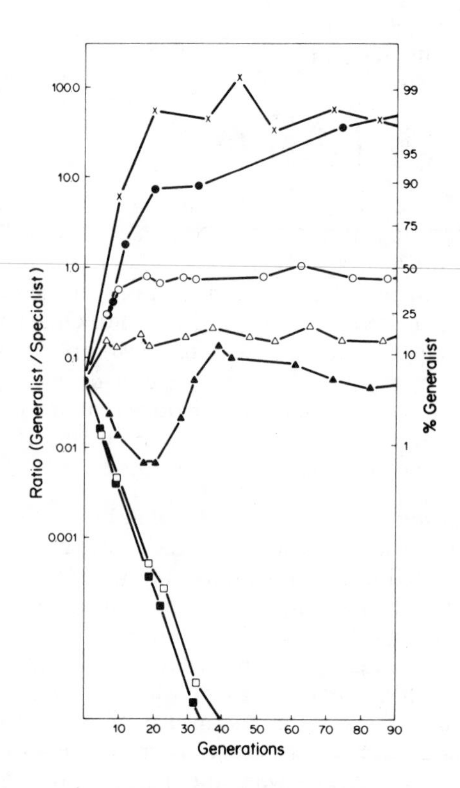

Fig. 5.13 Results of competition between an inefficient generalist strain of *E. coli* and an efficient specialist, with different percentages of lactose and maltose. ● 50% lactose; ○ 25% lactose; △ 10% lactose; ▲ 5% lactose; □ 1% lactose; ■ no lactose. (After Dykhuizen and Davies [70].)

5.4 Models of predation

The size of a predator population feeding on one prey is likely to depend on the size of its prey population. In terms of the logistic equation, the prey population determines the carrying capacity for the predators. The prey population must also vary according to the number eaten by the predators. How can we model such interactions?

The size of the predator population at time t is likely to be some function of the average size of the prey population during some period before t. It is commonly observed that population counts in a series of years are autocorrelated, population size in year t being correlated with that in year $t-1$, and so on. Therefore the prey population at time t is likely to be similar in size to the average prey population in years $t-n$ to $t+n$, where n is a small number. Thus the impact of predation on the prey population at time t might be proportional to the size of the prey population at time $t-T$. We can use equation (5.9) to model this situation; it predicts the impact of predators on the prey population, without requiring any explicit knowledge of the predator population.

We can regard herbivores as predators upon vegetation [71]. This allows us to model, for example, the impact of deer upon their range, and the effects of variations in the carrying capacity of the range upon the deer. This is an improvement upon the unrealistic assumption of a constant K in Section 5.2.2. In this model K is the level of the limiting resource defining the maximum sustainable population. The observed growth rate of K is assumed to depend partly on its maximum rate of growth R, partly on the extent to which current carrying capacity $K(t)$ is depressed below maximum carrying capacity K_0, and partly on the rate at which the population of deer removes vegetation (b per deer per unit time)

$$\frac{\mathrm{d}K}{\mathrm{d}t} = R[K_0 - K(t)] - bP(t) \tag{5.14}$$

We now ask a mathematician to solve equations (5.9) and (5.14). The result can be a stable population of grazers or predators, or else an oscillating one, depending upon the numbers that we substitute for the symbols. Heavy grazing or predation tends to result in oscillations.

Models can be based on quite unrealistic assumptions and yet give useful insights into population dynamics. The idea of one predator population feeding upon one prey population can have few parallels in the real world. Continued exponential growth limited only by predation is equally improbable, but is the basis of one of the most widely-quoted models of predator–prey interactions, often called the Lotka–Volterra model [72, 73].

In the absence of predators, we assume the prey population P_1 to be growing exponentially

$$\frac{\mathrm{d}P_1}{\mathrm{d}t} = aP_1 \tag{5.15}$$

In the absence of prey, the predator population P_2 shrinks exponentially

$$\frac{\mathrm{d}P_2}{\mathrm{d}t} = -gP_2 \tag{5.16}$$

The predator removes prey at a rate bP_1P_2 and so moderates the growth in equation (5.15)

$$\frac{\mathrm{d}P_1}{\mathrm{d}t} = aP_1 - bP_1P_2 \tag{5.17}$$

First, this implies that one predator removes prey at a rate bP_1, so the same fraction of the prey is removed by each predator in the same time, irrespective of P_1. *How realistic is this?*

Secondly, the predator population grows at a rate proportional to its rate of feeding YbP_1P_2, where Y is the yield constant similar to that in the chemostat model (Section 5.3). *How realistic is the assumption that the yield of predators is directly proportional to the predator population's food intake?*

Yb is constant, and if we call this c, then

$$\frac{dP_2}{dt} = cP_1P_2 - gP_2 \tag{5.18}$$

There are two possible outcomes to this model. For each set of values for the parameters a, b, c, and g there is a unique pair of population sizes of prey and predator where the rates of increase and decrease of the two populations cancel out. If these densities are attained, the populations remain at them indefinitely. Otherwise, the prey and predator populations oscillate with the same frequency but out of phase. The predator lags behind the prey and the oscillations continue with constant amplitude and frequency. Predator–prey oscillations with this general form have been observed in the laboratory (Fig. 5.14). This does not show, however, that the Lotka–Volterra model is an accurate reflection of the biological processes causing changes in numbers, because other models too can produce the same general pattern.

In Fig. 5.15 the abundance of the predatory larvae of the treehole mosquito *Toxorhynchites brevipalpis* lags behind the number of its prey (all pre-imaginal mosquitoes) in a fashion reminiscent of models for predator–prey cycles. Predator densities show a delayed density-dependent relationship with prey density, as would be expected for such a cycle. In fact, however, each increase in prey followed heavy rain, each decline a dry period. The lag ocurred because the predator recovered more slowly from its dry season depression than the prey. Despite the superficial similarity to Fig. 5.14, this was not a simple predator–prey oscillation.

If prediction is our goal, the best model for a particular situation is the one which makes the most precise forecasts of future population sizes. If understanding is our aim, the model which reflects the observed processes most faithfully would be preferred.

5.5 Further models

Throughout this chapter we have presented models for continuously-

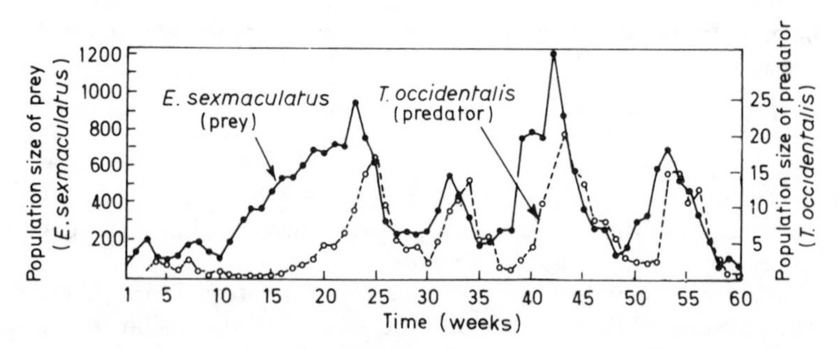

Fig. 5.14 Predator–prey cycle in the laboratory, of the general form predicted by the Lotka–Volterra model. The prey is an orange-eating mite, the predator a mite-eating mite. (After Huffaker, Shea and Herman [74].)

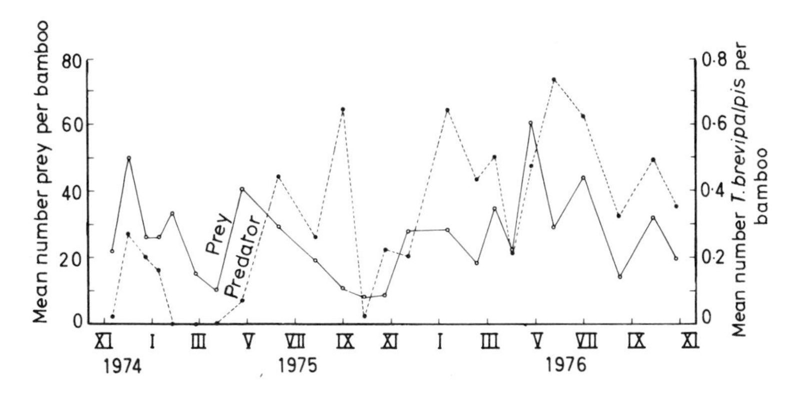

Fig. 5.15 Seasonal abundance of the predatory larvae of the treehole mosquito *Toxorhynchites brevipalpis* and its prey, in bamboo traps on the Kenya coast. (After Lounibos [75].)

growing populations, based on differential equations. This is quite unrealistic for animals which breed seasonally. An alternative approach is to use difference equations. These predict the size of generation $t+1$ from the sizes of generations t, $t-1$, $t-2$, ... $t-n$. Such models autom·tically incorporate time lags; general conclusions from them are similar to those from time-lagged logistic models.

Although we have ignored space in this chapter it is obvious to the field worker that the use of space by animals is a crucially important aspect of their population dynamics. Recent models and laboratory experiments confirm that the distribution of animals in space, and the structure of their environment, may play a large part in determining population changes (Chapter 7).

The mathematical models in this chapter were based on simple assumptions, yet led to complex patterns. It might be that the world is based on a few simple rules amenable to mathematical analysis. Alternatively, it might be necessary to create a new Earth in order to model this one adequately. Meanwhile, interactions between models and reality continue to provide fruitful new insights and generalizations.

6 Decision-making

Economic man classes animals as resources to be exploited or pests to be controlled. Applied population dynamics assists both these objectives.

6.1 Levels of exploitation

The basic question when deciding how many animals to kill is: how many can we take this season without reducing yields in seasons to come?

6.1.1 Self-regulated populations

It is commonly observed that breeding stocks of game-birds do not decrease when a fraction is removed by shooting each year. The game manager's job is to decide how big this fraction can safely be. At its simplest, he can regard a certain fraction of the mortality as due to overcrowding. Shooting reduces crowding and natural mortality. Each year the manager can decide how much mortality, due to crowding, would occur in the absence of shooting; he should then aim to shoot no more than this number of birds.

In the red grouse, birds that do not have territories are driven out of the population by the territory owners, which are more dominant. Fig. 6.1 shows how numbers continue to decline throughout the winter, well below the number left at the end of the shooting season. The fraction of the population that survives until the breeding season is already determined in autumn. Roughly half the birds in the August population take territories after the breeding season, and the number of territories determines next season's breeding stock. If a territory holder is shot, a previously non-territorial bird takes its place. The ultimate cause of

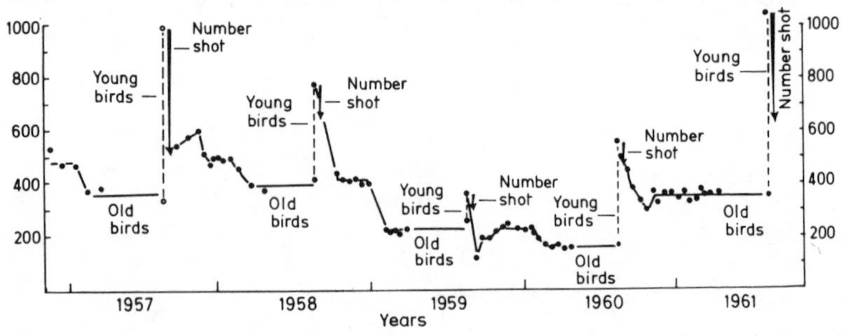

Fig. 6.1 Shooting mortality in a red grouse population. (After Jenkins, Watson and Miller [1].)

death is, for most birds in the non-territorial half of the population, simply the fact that they do not have territories. Whether the proximate cause of death is predation, disease, accident, parasites or shooting makes little difference.

If all non-territorial birds, or an equivalent number of non-territorial and territorial birds were shot, then a few territory owners would probably die over winter. Some mortality is unavoidable, or non-compensatory. Much mortality, however, is compensatory and would not occur were there territories enough for all the non-territorial birds to occupy before they died [76].

The biologist might point out that the shooting of birds at random may have different effects on the population than if an equivalent number of non-territorial birds had died as a result of the workings of the social system. In the short term it should damp the amplitude of fluctuations in breeding stock from year to year, because fluctuations are thought to depend on the nature as well as the number of birds in the population [64]. In the long term it might have unforeseen and possibly undesirable genetic effects.

6.1.2 Poorly regulated populations

Many deer populations fall into this category. It is not clear whether this is because they are *r*-selected animals adapted to exploiting particular stages in seral successions, or whether removal of predators by man has removed a regulatory mechanism (Section 4.4.2). The effect is the same. Unhunted populations may increase, starve and crash (Section 4.1.1). The deer manager aims to keep stocks at a level below what he regards as the carrying capacity of the range by setting hunting targets sufficient to do this.

He can determine when numbers are getting too high by looking at both range and deer. Overgrazed range is characterized by heavy, visible and measurable browsing on the favoured food plants. The animals' diet begins to include a higher proportion of less-favoured foods. The performance of the deer drops: females produce fewer young, males grow smaller antlers, fawn mortality is higher, growth rates are lower and all animals are smaller than on range which is not over-stocked. A proper cropping regime aims to keep animals below such densities. A second objective is to shoot all adults before they become senescent and die unused.

Once again, cropping by man is likely to have different effects from cropping by predators. One obvious aspect is that if trophy hunters shoot all the animals with big antlers, then there will be genetic selection against good trophies. Many managers are well aware of this and keep a high proportion of males with big antlers unshot until they have reached a good age.

Can you think of other effects that cropping by man may have?

6.1.3 Marine fish stocks

Estimating stocks and reproductive rates of game-birds and deer is a relatively simple matter. Marine biologists advising commercial fisheries have a far more difficult task.

The majority of commercial fish stocks occur within the relatively shallow nutrient-rich zones of sea overlying continental shelves. Separate stocks are divided by zones of deep ocean. This defines our populations. How are we to measure their abundance? The only practical answer is catch per unit effort.

When a virgin fishery is first exploited the catch per boat-day remains fairly constant for a while and then tends to decrease with increasing numbers of vessels. When fishing effort decreases, catch per unit effort increases again. Two unplanned experiments (the two world wars) showing this have occurred in the North Sea this century (Figs 6.2 and 6.3). It seems that heavy fishing limits fish stocks, but that stocks return to their former abundance when fishing is reduced. Sometimes, however, heavily-fished stocks decline to low levels and may fail to return to their previous levels even when fishing is stopped. The task of the

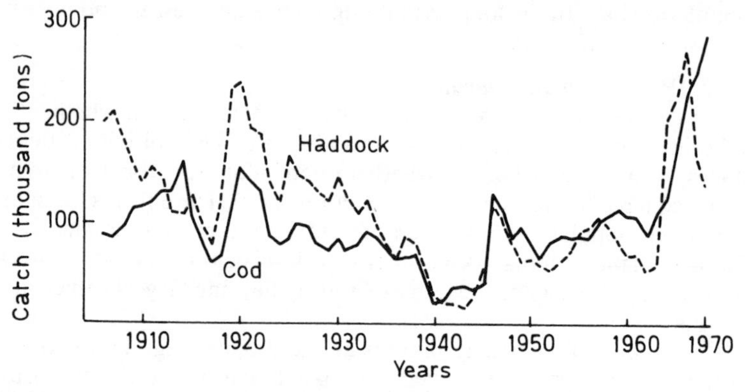

Fig. 6.2 Total international catches of cod and haddock in the North Sea, 1906–69. (After Gulland [8].)

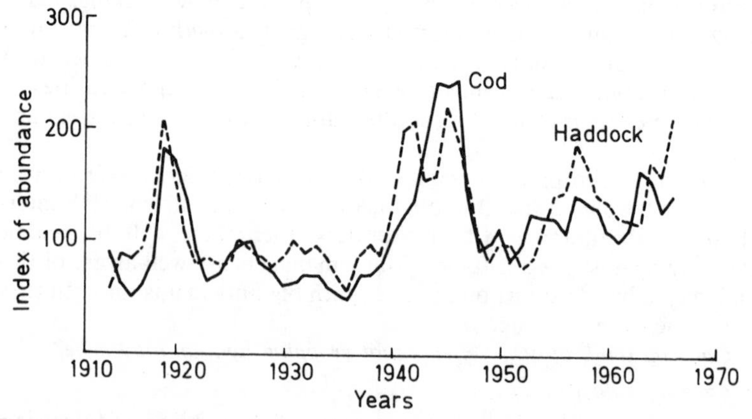

Fig. 6.3 Catch per unit time of cod and haddock in the North Sea by English and Scottish trawlers. (After Gulland [8].)

biologist is to make recommendations which will avoid overfishing and maximize the annual yield.

The major biological stumbling block is the difficulty in discovering what determines recruitment into the population of fishes big or old enough to be worth fishing. Most marine teleosts are very fecund, a female laying thousands or even millions of eggs each year. If an average fish laid 100 000 eggs per year, then a mortality rate of 99.999% from egg to recruit would ensure one recruit per adult per year, and a mortality rate of 99.998% would double this. It is not therefore surprising to find that the representation of different year classes within a population varies erratically (Fig. 3.5), presumably in response to variations in unknown mortality factors. It is, however, disconcerting that recruitment rates usually bear no relation whatsoever to the density of the parent stock [77].

One way of sidestepping this problem is to assume that the fishable population follows the logistic growth curve. Finding the 'maximum sustainable yield' (MSY in fishery jargon) is then simply a matter of maximizing the population growth rate, which occurs when the population is at one half of its maximum size (Fig. 5.7). Fig. 6.3 shows that practice did not depart far from this in the North Sea in the first half of this century. The idea behind this model is that increased reproduction, due to the population being below the carrying capacity K, can be removed by fishing. This is analogous with the equilibrium states obtained in Smith's work on water fleas (Section 5.2.1) and with bacterial populations in the chemostat (Section 5.3).

An alternative approach is to build a model which reflects the life-history of a typical fish stock (Fig. 6.4). Here, recruitment, growth of individuals and mortality are treated as separate processes which can vary separately. Growth is important because fish, like many cold-

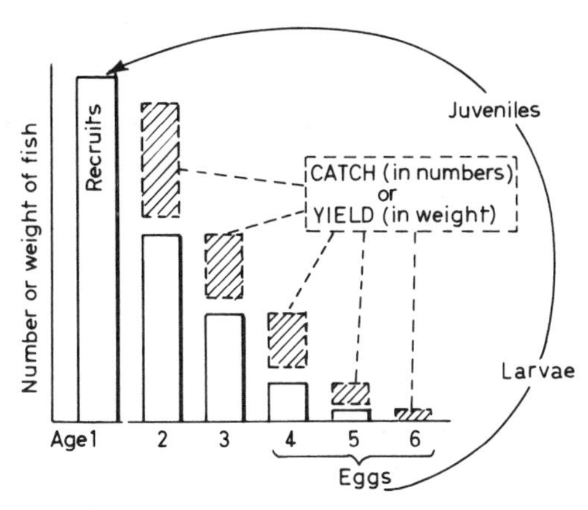

Fig. 6.4 The principle of the Beverton and Holt [78] model. The quantity of fish in year t is reduced by natural mortality and fishing in year $t+1$ and so on. Adult fish produce eggs, eggs hatch into larvae, larvae grow into juveniles, juveniles (often at one year of age) grow into recruits, and those recruits which reach maturity lay eggs. (After Jones [77].)

blooded animals, continue to grow in size throughout their life. The biomass of fish caught is more important than their numbers. Growth varies not only with age but also with population density, bigger fish of a given age occurring at lower densities.

The Beverton and Holt model (Fig. 6.4) can be considered in two halves: 'egg to recruit' and 'recruit to egg'. The recruit to egg half is of practical value because most of the parameters in it can be determined. Once the stock of the first age-class of recruits is known, then its contribution to stocks in subsequent seasons can be predicted. This is important because the 'strength' of different year-classes varies considerably (Fig. 3.5) and variations in this can affect both catches and quotas. The increased stock of haddock in the North Sea in the late 1960s (Fig. 6.3) was due to a succession of good year-classes, though why they were good is not known.

The form, if any, of the egg to recruit half remains the subject of intelligent guesswork. An important objective of any recommendations is to ensure that fishing effort never becomes so high that recruitment fails or the stock becomes extinct. This could happen if mortality in the early stages were of the inverse density-dependent type; for example, there might be so many predators that a certain minimum number of eggs is required before any survive to hatch. Such considerations favour the use of models which result in conservative quotas. On the other hand, a minor change in a mathematical model might result in drastic reductions in quotas and subsequent, possibly unnecessary, hardship for fishermen.

It is not known whether overfishing by itself can lead to a stock's extinction. Many heavily-fished stocks have declined to low levels and failed to recover when fishing was reduced; but in every case it has been possible to argue that a deterioration in the environment was a contributory factor. However, an annual catch equal to the maximum sustainable yield for average environmental conditions is bound to cause overfishing in those inevitable years when the environment is less favourable than average.

Theory suggests that maximizing total profits for the industry requires smaller catches than the maximum sustainable yield [77]. Indeed, a profitable fishing ground generally attracts new boats until the profit per boat drops to zero. Hence fishing effort tends to stabilize at a level greater than required either for the maximum sustainable yield or for the maximum economic yield (MEY). This is where the role of the biologist ends and that of the politician begins.

6.2 Controlling pests

The term 'pest' is a value-judgement and our toleration of economic competitors depends on aesthetic and political considerations as well as biological ones. Few voices are raised in defence of cereal aphids; but the control of bullfinches *Pyrrhula pyrrhula* eating the buds of fruit trees and grey seals *Halichoerus grypus* eating fish are much more emotive issues.

However, the control of such slowly-reproducing vertebrates is a relatively easy technical problem. More spectacular, and technically more difficult to deal with, are outbreaks of fast-reproducing insects.

6.2.1 Self-regulated outbreaks

The periodical cicada *Magicicada* spp. is not much of a pest, but it illustrates an important principle. The nymphs in any one population suck juices from the roots of forest or suburban trees and emerge from the ground, become adults, mate, lay their eggs and die, all within the same few weeks of every 17th (or, in part of their range, 13th) year [79]. There are three separate species; and, in the same area, they always do it together. When they emerge, they reach densities far in excess of any of the more conventional cicadas. The obvious conclusion is that these outbreaks are genetically programmed.

A result of this extraordinary life-cycle is that they escape limitation by predators. The only organism specialized to feed on them is the fungus disease *Massospora cicadina*; even this is not very effective at controlling numbers [80]. Similar life-histories are also well known in plants. Many bamboo populations seed synchronously at regular, long intervals from 3–120 years [81]. This allows them to produce vast quantities of seed from which young plants grow despite heavy predation by rats, swine, elephants, rhinoceroses, jungle fowl, pheasants, humans and others. The theory of self-regulated oscillations in voles and other cyclic vertebrates (Section 4.4.3) might also be interpreted in this light: perhaps such cycles minimize the effects of predation on the population.

Entomologists, however, have been more concerned with analysing the environmental conditions which lead to outbreaks of insect pests.

6.2.2 Weather and pest outbreaks

A swarm of locusts, weighing perhaps thousands of tons, appears out of thin air, descends on crops and eats its own weight in a day. Much work has been done on the ten species of swarming locusts.

The origin of locusts was a mystery for thousands of years. In 1921 Uvarov [82] pointed out that locusts were the dispersing phase of certain solitary grasshoppers. When favourable weather encourages breeding, the grasshoppers increase in density. High densities cause individuals to aggregate in groups; frequent contact with each other excites changes in behaviour, physiology, colour and morphology which culminate in the emigration of a swarm of locusts (Fig. 6.5).

Rainfall seems to be the crucial factor preceding swarm formation. In species which live in arid areas swarms occur after heavy rains, and after droughts in the species of wet or swampy areas. Swarms may originate in relatively small 'outbreak areas' and travel hundreds or even thousands of miles downwind into much bigger 'invasion areas'.

Why unusual amounts of rain are followed by outbreaks is not clear. White [84] has put forward the idea that many herbivores are limited by

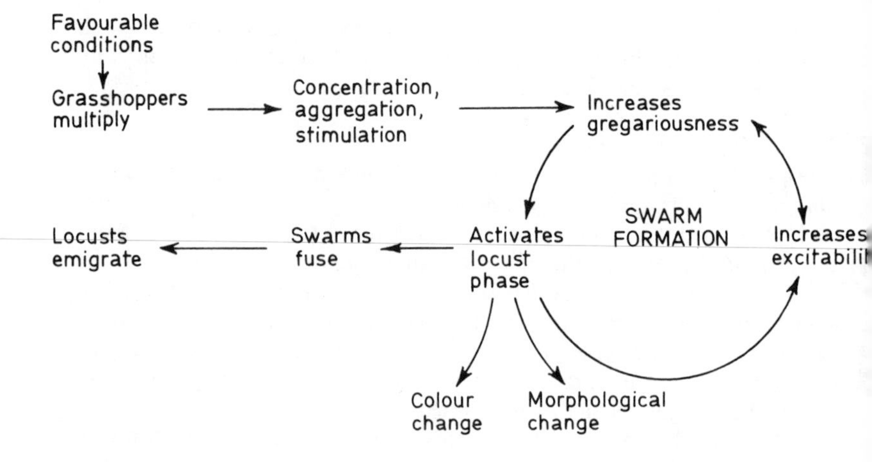

Fig. 6.5 Mechanism of swarm formation in locusts. (After Key [83].) See [2] for a more detailed account of locust ecology.

a shortage of nitrogenous food for their very young. He points out that plants which are stressed either by excessive rainfall or by unusual drought often contain unusually high concentrations of nitrogen in their tissues. When this happens, the herbivores breed faster. Outbreaks may occur when large areas of vegetation are stressed by the weather or other factors. A cause of outbreaks of defoliating insects on plants along motorways seems to be the plants' high nitrogen content, due to oxides of nitrogen from car exhaust fumes [85].

White's idea is not inconsistent with a role for the genetic programming of outbreaks. An animal may break out of its normal limits of abundance at those times when the environment allows it to reproduce fastest. Neither of these ideas is widely accepted. *What do you think are the reasons for this?*

Irrespective of biological mechanisms, the knowledge that rainfall is correlated with outbreaks is valuable to control programmes. A network of observers report aggregating swarms, and knowledge of the weather conditions that precede outbreaks is used to identify potential outbreak areas. If such areas are reasonably compact, they can then be sprayed with insecticides.

6.2.3 The numerical approach

Predators may keep a pest in check for many years, with only occasional outbreaks. The impulse to spray outbreaks of insect pests with poisons should be tempered by consideration of possible effects on the predators.

The spruce budworm *Choristoneura fumiferana* irrupts every 40 years or so, causing massive defoliation of Canadian coniferous forests. The dynamics of this population have been described [71] in terms of the simple logistic equation (5.6), modified by subtracting a term designed to mimic the impact of predators on the budworm

$$\frac{dP}{dt} = rP\left(1 - \frac{P}{K}\right) - bP'\left(\frac{P^2}{P_S^2 + P^2}\right) \tag{6.1}$$

where P is the budworm population, K is the carrying capacity and is related to the leaf area of the forest, P' is the predator population, and P_S the population of budworms at which the predator attack rate maximizes at a rate of b budworms per predator. If the rate of attack had been constant the last term would have been simply bP'; the extra term in brackets models the idea that predators are likely to be eating many other prey until budworm densities reach P_S.

We can represent equation (6.1) graphically by starting with Fig. 5.7 and regarding this as the logistic growth rate of a budworm population without predators. If we now superimpose the predation term as a dashed line, we get Fig. 6.6. At the three 'equilibrium' points where the predation curve crosses the logistic curve, predators are eating budworms as fast as they can reproduce and keeping the budworms' population size constant. Two of these equilibria are stable: at each, any slight change in the budworm population P is followed by a faster change in the predation rate and this returns P to the equilibrium point. The middle equilibrium point is unstable: any change in the budworm population is accompanied by a change in predation which helps to shift the population towards one of the stable equilibria. The system will tend to move towards one of the stable equilibria from wherever it starts.

It is assumed that the lower equilibrium population A of budworms has little impact on the forest, and that K gradually increases as the forest grows over several decades. Hence the budworm population too increases, slowly. At some point it is supposed that the budworm population is no longer limited by predation and irrupts to the higher

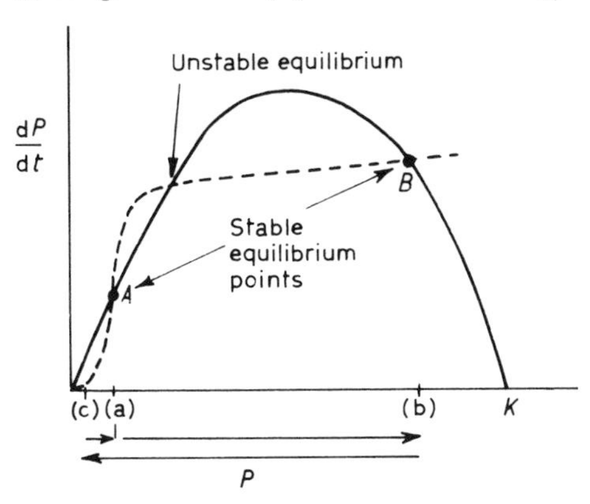

Fig. 6.6 Model of the budworm cycle: (a) P starts at A; (b) P irrupts to B; this depresses K, so P drops to low levels and then (c) returns to A as K recovers.

equilibrium population B. This higher population rapidly defoliates the forest, decreasing K and so P. The ensuing low population is once more controllable by predators and so the cycle begins again.

What are the untested biological assumptions in this model? Might there be alternative biological explanations for the irruptions?

The model provides some interesting insights. It depends on P_s being much lower than K: the regulating factor, thought to be predation, must become ineffective at fairly low densities of budworms (at A). Application of insecticide might help to limit densities: detailed analysis suggests that this would lead to a single equilibrium point at a density intermediate between A and B. To keep this under control would require continued and uneconomical use of insecticide.

Experience tends to confirm this prediction. An increasing opinion in Canada seems to be that it is best to let the budworm irrupt and starve. Many managers, however, have yet to learn this lesson.

It is generally thought that many insect pests in unmanaged environments are regulated by their predators. Predators generally have an intrinsic rate of increase slower than that of their prey. Hence, irrespective of detailed analysis, it is obvious that pesticides are likely to damage the predator populations more than their prey. Pesticides can be a drug; once hooked, the manager cannot kick the habit because he can no longer rely on the help of the predators.

7 Time, space and chance

When studying a small population in a defined area, we attempt to explain the temporal pattern of its changes in numbers in terms of gains and losses. Most of this book has taken this approach, which has its limitations. Here we end by indicating some of them. An important question is whether a much larger population comprising, say, the entire stock of one species, is limited by the same factors. Is our microcosm a faithful miniature of the macrocosm, or does the latter have properties of its own?

7.1 Distribution and population regulation

Periodical cicadas show remarkable and extreme temporal patterns of abundance (Section 6.2.1). Their spatial pattern, the way in which individual populations are distributed within the available habitat, is less unusual. During each cycle, many areas of apparently suitable habitat remain unused [80].

Fig. 7.1 shows annual changes in distribution of the elder aphid *Aphis sambuci*. Elder aphid has winter host plants *Sambucus* and summer hosts *Rumex* in every part of Great Britain, but its own distribution

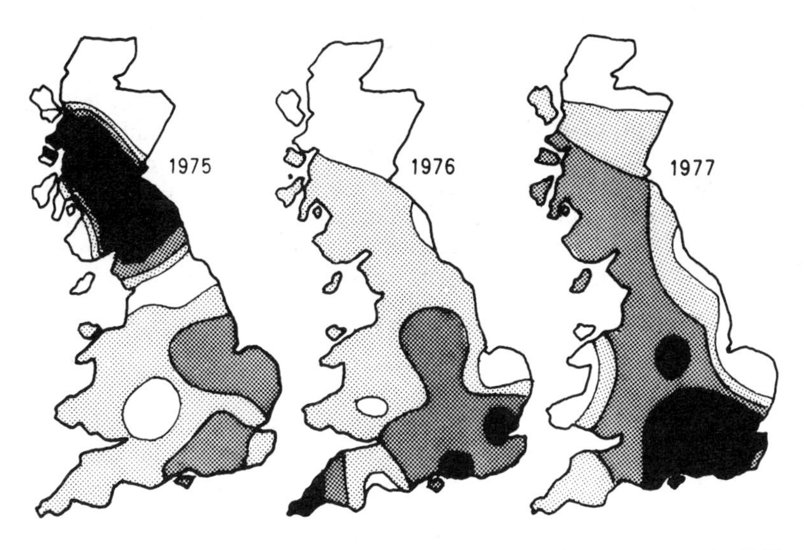

Fig. 7.1 Distribution of the elder aphid in Great Britain. (After Taylor and Taylor [86].)

differs each year [86]. There is no permanent population centre. The British population is clearly well below its carrying capacity.

Clearly, the elder aphid survives by dispersal (Section 3.3.1); but the observed distribution cannot be explained simply in terms of patches of habitat appearing and disappearing as stages in seral successions. Resource limitation is an inadequate explanation of what we see. The limiting factor is the inability of the individual to find the resources: its dispersal and the behaviour which controls it.

This seeming incompetence might be of advantage to the aphid. Populations of the periodical cicada which are newly-established on fresh ground often escape infection by the *Massospora* fungus for some time [80]. In general, populations newly-established in areas that have been vacant for some time often flourish until their parasites, diseases and predators catch up with them. Whether by incompetence or natural selection, the elder aphid may have hit upon a good survival strategy. Any full account of its population dynamics must acknowledge the role of chance or random events.

7.2 Habitat structure and population regulation

Individuals require several resources to survive. A heterogeneous habitat is likely to provide all the necessities of life in a smaller space than a homogeneous one. It is often observed that herbivorous animals are denser in areas where there are many small patches of differently-structured vegetation: more edge, more herbivores. At this level, spatial heterogeneity may be regarded as one factor potentially limiting a population.

Snowshoe hares in parts of North America oscillate in density about once every ten years [27]. At high densities, they browse their food heavily, and then, while suffering heavy predation, decline to very low densities. As they decline, so their distribution alters. At the trough of the cycle in Alaska, they desert open habitat and are seen only in dense thickets of spruce or willow and alder [87]. As numbers build up once more, they re-colonize the open habitat. Evidently, the population dynamics of hares in open areas and refuges are quite different.

Refuges may differ physically from less favourable areas. Or, they may simply be places where chance allows populations to survive. In simple laboratory experiments, a predator may rapidly exterminate its prey. To avoid this, and achieve the predator–prey oscillation of Fig. 5.14, Huffaker, Shea and Herman [74] constructed a complex of 252 oranges with vaseline barriers. This allowed some local populations of the prey to avoid predators for long enough to found new populations in a deadly game of hide-and-seek. The real world is far more complex than 252 oranges, and the chances of survival by dispersal (Section 3.3.1) may be improved by environmental heterogeneity.

Whether or not a herbivore cycles in abundance may depend on the structure of its habitat [57, 88]. The key observation, for grouse and ptarmigan, snowshoe hares and voles *Microtus* spp. is that cycles occur only in large tracts of uninterrupted habitat. In areas where the habitat is patchy, where 'islands' of suitable habitat are surrounded by a 'sea' of unsuitable habitat, cycles do not occur. A reasonable suggestion is that populations in patches of good habitat are regulated by constant emigration into the surrounding sink of poor habitat. On continuous tracts of good habitat, one population's emigration (Fig. 4.7) is another's immigration. It might be that immigration plays a necessary role in triggering those demographic processes which cause cyclic declines in numbers. Just how this might work is not clear.

7.3 The future

Population dynamics is a living discipline. We hope that what we have written today, you will disprove tomorrow.

Acknowledgement

We thank Professor Mark Williamson for reading the typescript and making comments.

References

[1] Jenkins, D., Watson, A. and Miller, G.R. (1963), *J. Anim. Ecol.,* **32**, 317–376.

[2] Krebs, C.J. (1978), *Ecology: The Experimental Analysis of Distribution and Abundance* (Second Edition), Harper and Row, New York.
In this excellent text you will find more detailed discussions of many points, including the history of many of the concepts and people in population dynamics, that we have dealt with only briefly.

[3] Strandgaard, H. (1972), *Danish Review of Game Biology*, **7**, 1–205.

[4] Elliott, S.M. (1977), *Some Methods for the Statistical Analysis of Samples of Benthic Invertebrates* (Second Edition), Freshwater Biological Association, Ferry House, Ambleside, Cumbria, U.K.
A valuable guide to the main types of statistical distribution shown by animals, written by a biologist for biologists.

[5] Harris, M.P. and Murray, S. (1981), *Bird Study*, **28**, 15–20.

[6] Betts, M.M. (1955), *J. Anim. Ecol.,* **24**, 282–323.

[7] Lockie, J.D. (1964), *J. Anim. Ecol.,* **33**, 349–356.

[8] Gulland, J.A. (1971), in *Dynamics of Populations* (den Boer, P.J. and Gradwell, G.R., eds), pp. 450–468, Centre for Agricultural Publishing and Documentation, Wageningen, Netherlands.
A good account of the effects of exploitation upon marine animals. Many of the papers in this book are well worth reading.

[9] Southern, H.N. (1970), *J. Zool.*, Lond., **162**, 197–285.

[10] Wilson, D.S. (1980), *The Natural Selection of Populations and Communities*, Benjamin/Cummings, Menlo Park, California.
A new model of group selection.

[11] Andersen, J. (1957), *Danish Review of Game Biology*, **3**, 85–131.

[12] Tagore, Rabindranath (1976), *Fireflies* (Second Edition), Collier, Macmillan, New York.
Brief poems, not a text on fireflies.

[13] Pearl, R. (1928), *The Rate of Living*, Knopf, New York.

[14] Lowe, V.P.W. (1969), *J. Anim. Ecol.,* **38**, 425–457.

[15] Andersen, J. (1953), *Danish Review of Game Biology*, **2**, 127–155.

[16] Sunada, J.S. (1976), *Calif. Fish and Game*, **62**, 213–224.

[17] Romanov, A.N. (1979), *The Capercaillie*, Nauka, U.S.S.R. (*in Russian*).

[18] Varley, G.C. and Gradwell, G.R. (1968), *Symp. Roy. Entomol. Soc.*, London, **4**, 132–142.

[19] Manly, B.F.J. (1977), *Oecologia*, **31**, 111–117.
Points out some pitfalls in k-factor analysis.

[20] Wilbur, S.R. (1980), *Calif. Fish and Game*, **66**, 40–48.

[21] Connell, J.H. (1979), in *Population Dynamics* (Anderson, R.M., Turner, B.D. and Taylor, L.R., eds), pp. 141–163, Blackwell Scientific Publications, Oxford.
Overturns the long-held view that ecological diversity in the tropics is the result of stable conditions. The book as a whole reflects the current emphasis on numerical theories in population dynamics.

[22] Blau, W.S. (1980), *Ecology*, **61**, 1005–1012.

[23] Allen, D.L. (1979), *Wolves of Minong*, Houghton Mifflin, Boston.
A personal account of wolves and moose on Isle Royale, with many references.

[24] Whittaker, J.B. (1971), *J. Anim. Ecol.,* **40**, 425–443.
[25] Watson, A. and Moss, R. (1970), in *Animal Populations in Relation to their Food Resources* (Watson, A., ed.), pp. 167–220, Blackwell Scientific Publications, Oxford.
[26] Wiegert, R.G. (1964), *Ecol. Monogr.*, **34**, 217–241.
[27] Keith, L.B. and Windberg, L.A. (1978), *Wildlife Monographs*, No. 58.
[28] Solomon, M.E. (1964), *Adv. Ecol. Res.*, **2**, 1–58.
A useful review of numerical methods for detecting and measuring regulation (in its narrow sense) in insect populations.
[29] Klein, D. (1968), *J. Wildl. Mgmt.*, **32**, 350–367.
[30] Leopold, A.S., Riney, T., McCain, R. and Tevis, L., Jr. (1951), *Game Bulletin* No. 4, Calif. Dept. Fish and Game.
[31] Dempster, J.P. and Lakhani, K. (1979), *J. Anim. Ecol.,* **48**, 143–163.
[32] Savory, C.J. (1978), *J. Anim. Ecol.,* **47**, 269–282.
[33] McGowan, J.D. (1969), *Auk*, **86**, 142–143.
[34] Leopold, A.S. (1977), *The California Quail*, University of California, Berkeley.
[35] Gallizioli, S. (1965), *Quail Research in Arizona*, Arizona Game and Fish Dept., Phoenix. Quoted by [34].
[36] Telfer, E.S. (1967), *J. Wildl. Mgmt.*, **31**, 418–425.
[37] Duncan, J.S., Reid, H.W., Moss, R., Phillips, J.P.D. and Watson, A. (1978), *J. Wildl. Mgmt.*, **42**, 500–505.
[38] Chapman, R.F. and Page, W.W. (1979), *J. Anim. Ecol.,* **48**, 271–288.
[39] Pearson, O.P. (1964), *J. Mamm.,* **45**, 177–188.
[40] Estes, J.A., Smith, N.S. and Palmisano, J.F. (1978), *Ecology*, **59**, 822–833.
[41] Snyder, W.D. (1967), *Tech. Publs Colo. Game Fish Pks Dep.*, No. 19.
[42] Saunders, J.W. and Smith, M.W. (1962), *Trans. Am. Fish. Soc.,* **91**, 185–188.
[43] Bustard, H.R. (1969), *J. Anim. Ecol.,* **38**, 35–51.
[44] Houston, D.C. (1977), *J. Appl. Ecol.,* **14**, 1–15.
[45] Charles, J. (1972), Territorial Behaviour and the Limitation of Population Size in the Crow *Corvus corone* and *Corvus cornix*. Unpublished Ph.D. thesis, University of Aberdeen.
[46] Yom-Tov, Y. (1974), *J. Anim. Ecol.,* **43**, 479–498.
[47] Pearson, O.P. (1971), *J. Mamm.,* **52**, 41–49.
[48] Krebs, C.J. (1966), *Ecol. Monogr.,* **36**, 239–273.
[49] Lidicker, W.Z. (1978), in *Populations of Small Mammals under Natural Conditions* (Snyder, D.P., ed.), pp. 122–141, Pymatuning Laboratory of Ecology, University of Pittsburgh.
Puts forward a holistic view of population dynamics.
[50] Lack, D. (1969), *J. Anim. Ecol.,* **38**, 211–218.
[51] Watson, A. and Moss, R. (1979), *Orn. Fenn.,* **56**, 87–109.
Reviews population cycles in grouse and ptarmigan.
[52] Watson, A. and Parr, R.A. (1981), *Orn. Scand.,* **12**, 55–61.
[53] Watson, A. and Moss, R. (1980), *Ardea*, **68**, 103–111.
[54] Thompson, D.C. (1978), *Ecology*, **59**, 708–715.
[55] Krebs, C.J., Keller, B.L. and Tamarin, R.H. (1969), *Ecology*, **50**, 587–607.
[56] Tamarin, R.H. (1977), *Ecology*, **58**, 1044–1054.
[57] Leuze, C. (1976), Social Behaviour and Dispersion in the Water Vole, *Arvicola terrestris*, Lacépède. Unpublished Ph.D. thesis, University of Aberdeen.
[58] Jenkins, D., Watson, A. and Miller, G.R. (1964), *J. Appl. Ecol.,* **1**, 183–195.

[59] Petrusewicz, K. (1966), *Ekol. Pol.* A, **14**, 413–436.
One example of the well-documented fact that laboratory populations started with the same number of animals reach different asymptotes.
[60] Redfield, J.A. (1974), *Evolution*, **27**, 576–592.
[61] Myers, J.H. and Krebs, C.J. (1971), *Ecol. Monogr.*, **41**, 53–78.
[62] Greenwood, P.J., Harvey, P.H. and Perrins, C.M. (1979), *J. Anim. Ecol.*, **48**, 123–142.
[63] Chitty, D. (1967), *Ecol. Soc. Austr., Proc.*, **2**, 51–78.
Here is put forward the idea, now well accepted, that genetic changes in a population can occur quickly enough to affect population dynamics in a few generations. Whether they actually do so often enough to be important is still moot.
[64] Moss, R. and Watson, A. (1980), *Ardea*, **68**, 113–119.
[65] Wynne-Edwards, V.C. (1962), *Animal Dispersion in Relation to Social Behaviour*, Oliver and Boyd, Edinburgh.
[66] Smith, F.E. (1963), *Ecology*, **44**, 651–663.
[67] May, R.M. (1976), in *Theoretical Ecology* (May, R.M., ed.), pp. 4–25, Blackwell Scientific Publications, Oxford.
This book gives a readable introduction to many evolving ideas in theoretical ecology.
[68] Southwood, T.R.E. (1976), in *Theoretical Ecology* (see [67]), pp. 26–48.
An exposition of the r-selection/K-selection viewpoint on animal life histories.
[69] Herbert, D., Elsworth, R. and Telling, R.C. (1956), *J. Gen. Microbiol.*, **44**, 601–622.
[70] Dykhuizen, D. and Davies, M. (1980), *Ecology*, **61**, 1213–1227.
[71] May, R.M. (1977), *Nature*, **269**, 471–477.
Reviews, in simple form, a variety of mathematical models for ecological systems. The models show how continuous variation in a causal factor can have discontinuous effects.
[72] Lotka, A.J. (1925), *Elements of Physical Biology* (Reprinted 1956 by Dover Publications, New York).
[73] Volterra, V. (1926), *Nature*, **118**, 558–560.
[74] Huffaker, C.B., Shea, K.P. and Herman, S.G. (1963), *Hilgardia*, **34**, 305–330.
[75] Lounibos, L.P. (1979), *J. Anim. Ecol.*, **48**, 213–236.
[76] Watson, A. and Moss, R. (1972), *Proc. Int. Orn. Congr.*, **15**, 134–149.
[77] Jones, R. (1979), *Investigación Pesquera*, **43**, 1–20.
A lucid summary of theory in fish population dynamics.
[78] Beverton, R.J.H. and Holt, S.J. (1957), *Fish. Inv. Ser. II*, **19**.
[79] Lloyd, M. and Dybas, H.S. (1966), *Evolution*, **20**, 466–505.
[80] White, J., Lloyd, M. and Zar, J.H. (1979), *Ecology*, **60**, 305–315.
[81] Janzen, D.H. (1976), *Annu. Rev. Ecol. Syst.*, **7**, 347–391.
[82] Uvarov, B.P. (1921), *Bull. Entomol. Res.*, **12**, 135–163.
[83] Key, K.H.L. (1950), *Quart. Rev. Biol.*, **25**, 363–407.
[84] White, T.C.R. (1976), *Oecologia*, **22**, 119–134.
[85] Port, G.R. and Thompson, J.R. (1980), *J. Appl. Ecol.*, **17**, 649–656.
[86] Taylor, R.A.J. and Taylor, L.R. (1979), in *Population Dynamics* (see [21]), pp. 1–27.
[87] Wolff, J.O. (1980), *Ecol. Monogr.*, **50**, 111–130.
[88] Abramsky, Z. and Tracy, C.R. (1979), *Ecology*, **60**, 349–361.

Index